175 Topics in Current Chemistry

W0245934

Supramolecular Chemistry II - Host Design and Molecular Recognition

Editor: E. Weber

With contributions by
A. Galán, O. Hayashida, J. Kikuchi,
U. Lüning, J. de Mendoza, W. L. Mock,
Y. Murakami, R. J. M. Nolte, C. Seel,
R. P. Sijbesma

With 44 Figures and 18 Tables

 Springer

This series presents critical reviews of the present position and future trends in modern chemical research. It is addressed to all research and industrial chemists who wish to keep abreast of advances in their subject.

As a rule, contributions are specially commissioned. The editors and publishers will, however, always be pleased to receive suggestions and supplementary information. Papers are accepted for "Topics in Current Chemistry" in English.

ISBN 978-3-662-14861-7   ISBN 978-3-540-49106-4 (eBook)
DOI 10.1007/978-3-540-49106-4

Library of Congress Catalog Card Number 74-644622

© Springer-Verlag Berlin Heidelberg 1995
Originally published by Springer-Verlag Berlin Heidelberg New York in 1995
Softcover reprint of the hardcover 1st edition 1995

Typesetting: Macmillan India Ltd., Bangalore-25
SPIN: 10477233   51/3020 - 5 4 3 2 1 0 - Printed on acid-free paper

Attention
all "Topics in Current Chemistry" readers:

A file with the complete volume indexes Vols.22 (1972) through 174 (1995) in delimited ASCII format is available for downloading at no charge from the Springer EARN mailbox. Delimited ASCII format can be imported into most databanks.

The file has been compressed using the popular shareware program "PKZIP" (Trademark of PKware Inc., PKZIP is available from most BBS and shareware distributors).

This file is distributed without any expressed or implied warranty.

To receive this file send an e-mail message to:
SVSERV@VAX.NTP. SPRINGER.DE
The message must be:"GET/CHEMISTRY/TCC_CONT.ZIP".

SVSERV is an automatic data distribution system. It responds to your message. The following commands are available:

HELP	returns a detailed instruction set for the use of SVSERV
DIR (name)	returns a list of files available in the directory "name",
INDEX (name)	same as "DIR",
CD <name>	changes to directory "name",
SEND <filename>	invokes a message with the file "filename",
GET <filename>	same as "SEND".

For more information send a message to:
INTERNET:STUMPE@SPINT. COMPUSERVE.COM

Preface

It is now common knowledge: 'Supramolecular Science' with 'Supramolecular Chemistry' being its dominant part has a great future. In this field, scientists with organic, inorganic and theoretical backgrounds have produced novel compounds that show remarkably selective chemical behaviour including ion and molecular separations, transport and catalysis, up to enzyme-like action, controlled functionally and molecular intelligence. This work has attracted attention in many fields of application where selectivity in general and programmed supramolecular functions are of prime concern including chemical devices based on sensing, switching and transformation, nanotechnology and other high-tech uses. All these processes and phenomena imply a particular property, i.e. 'molecular recognition' - the key word and leading idea of the present book.

The creation of molecules which have this property is not a trivial concern. It is obvious from the five indiviual chapters covered in the book, there are outstanding experts with a master's touch in the field who have constructed molecular holes, niches, cavities and clefts that are capable of forming selective host-guest complexes and inclusion compounds.

Although this book is the second volume published under the topic 'Supramolecular Chemistry', it is more than merley a sequal to the previous volume ('Supramolecular Chemistry I - Directed Synthesis and Molecular Recognition', Top. Curr. Chem., Vol. 165) since it introduces new important aspects of supramolecular receptor design. This is greatly acknowledged by the editor. He wishes to express his heartfelt appreciation to all the contributers who have made this book possible.

Freiberg, December 1994   Edwin Weber

Table of Contents

Table of Contents of Volume 165

Cucurbituril

William L. Mock

Department of Chemistry, University of Illinois at Chicago, Chicago, IL 60607-7061, USA

Table of Contents

Cucurbituril is a novel nonadecacylic cage compound with an exceptional capacity to encapsulate alkylammonium ions within its hollow core. The structure readily self-assembles from acidic condensations between urea, glyoxal and formaldehyde. As a receptor for substituted ammonium ions, it demonstrates molecular recognition phenomena with high specificity, allowing quantitative thermodynamic evaluation of hydrophobic effects, as well as permitting systematic examination of the dynamics of noncovalent binding. Cucurbituril also induces an azide-alkyne 1,3-dipolar cycloaddition, providing a kinetic acceleration of approximately 10^5-fold, and in the process exemplifying the Pauling principle of catalysis.

1 Introduction

It may come as a surprise to learn that one of the more selective and most easily accessible of the synthetic molecular receptors was first reported as long ago as 1905. In that year Behrend [1] described the acidic condensation between an excess of formaldehyde (CH_2O) and the bis-ureide glycoluril ($C_4H_6N_4O_2$, itself a condensation product of CHOCHO and two NH_2CONH_2). The initial product so obtained should probably be regarded as a crosslinked, aminal-type polymer by virtue of its physical properties (amorphous character, insolubility in all common solvents). In seeking a more tractable material from this precipitate, Behrend and his coworkers resorted to treatment with hot, concentrated sulfuric acid, which eventually dissolves the substance. When such solution is diluted with cold water and subsequently boiled, a crystalline precipitate is obtained, Eq. (1). Methods of the time were not adequate for identifying the product, which was simply characterized as $C_{10}H_{11}N_7O_4 \cdot H_2O$. The substance was shown to be exceedingly stable toward a number of potent reagents. Also, a series of crystalline complexes incorporating a surprising variety of metal salts and dye stuffs was recorded.

$$\text{"glycoluril"} + \text{excess } CH_2O \xrightarrow{H_2O, HCl} \text{precipitate} \xrightarrow[110\ °C]{\text{concd } H_2SO_4}$$

$$\xrightarrow[0\text{-}10\ °C]{H_2O} \text{solution} \xrightarrow{\Delta} \text{"cucurbituril"} \tag{1}$$

In the early 1980s we came across this report and were intrigued by it. The preparation was repeated without difficulty and a spectral characterization of the product was carried out. It was immediately apparent that the imidazolone ring of the glycoluril remained intact (IR, $1720\ cm^{-1}$) and the NMR spectrum, consisting of only three equal-intensity signals, indicated a highly symmetrical, non-aromatic structure: δ 5.75 (singlet, from glycoluril nucleus), 4.43 and 5.97 (doublets, formal-derived). The inferred reaction stoichiometry was then

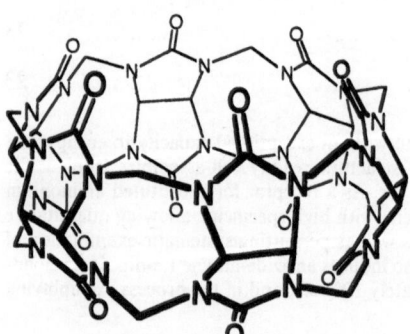

Fig. 1. Perspective representation of the structure of cucurbituril; from [15] with permission

$nC_4H_6N_4O_2 + 2nCH_2O \rightarrow (C_6H_6N_4O_2)_n + 2nH_2O$. That prompted an X-ray crystallographic investigation [2, 3], revealing the hexameric macropolycyclic structure of composition $C_{36}H_{36}N_{24}O_{12}$ now known as *cucurbituril* (Fig. 1). The designated trivial name (pronunciation: kyu · ker'· bit · yur · eel') derives tongue-in-cheek from the general resemblance of models of the molecule to a gourd or a pumpkin (of botanical family Cucurbitaceae).[1]

A remarkable chemical aspect of the structure is that all of its 19 rings are held together entirely by *aminal* linkages, formed of the constituents formaldehyde, glyoxal, and urea. The most notable feature as regards molecular recognition is the presence of an internal cavity of approximately 5.5-Å diameter within the relatively rigid macrocyclic structure, to which access is provided by a 4-Å diameter "occulus" situated among the carbonyl groups on both the top and the bottom of the molecule as depicted. While some developmental work on the synthesis has been carried out, the procedure of Behrend has not been substantially improved upon. We suspect that the material ultimately obtained is the product of an acid-induced, thermodynamically controlled rearrangement of an initially formed, irregular, macromolecular condensation product. In view of the subsequently elaborated cation-binding properties possessed by the carbonyl-fringed "occuli", we think it likely that a *template* synthesis is involved, with hydronium ions at this locus providing nuclei for assembly of the convex structure. That would explain why only the cyclic *hexamer* is produced; oligomers with greater or fewer glycoluril units have not been detected in the standard synthesis. It might be noted that a cyclic *pentamer* has been secured under milder acidic conditions with formaldehyde and dimethylglycoluril (i.e. the bis-ureide obtained from 2,3-butanedione rather than glyoxal) [4, 5]. This observation may indicate a subtle oligomerization dependence contingent upon small bond angle deformations within the monomer units. The methylated derivative has a smaller cavity, and has not provided any host-guest chemistry, aside from a possible complexation with acetylene.

2 Molecular Recognition

2.1 General Considerations

The cavity inside cucurbituril can hold small organic molecules. This has been established crystallographically [3], and is easily investigated in solution by

---

[1] The proper (current Chemical Abstracts index) name is dodecahydro-1H, 4H, 14H, 17H-2, 16:3, 15-dimethano-5H, 6H, 7H, 8H, 9H, 10H, 11H, 12H, 13H, 18H, 19H, 20H, 21H, 22H, 23H, 24H, 25H, 26H-2, 3, 4a, 5a, 6a, 7a, 8a, 9a, 10a, 11a, 12a, 13a, 15, 16, 17a, 18a, 19a, 20a, 21a, 22a, 23a, 24a, 25a, 26a-tetracosaazabispentaleno[1''', 6''':5'', 6'', 7'']cycloocta[1'', 2'', 3'':3', 4']pentaleno (1', 6':5, 6, 7)-cycloocta(1, 2, 3-gh:1', 2', 3'-g'h')cycloocta(1, 2, 3-cd:5, 6, 7-c'd')dipentalene-1, 4, 6, 8, 10, 12, 14, 17, 19, 21, 23, 25-dodecone. Alternatively, it is 1, 3, 5, 7, 10, 12, 14, 16, 19, 21, 23, 25, 28, 30, 32, 34, 37, 39, 41, 43, 46, 50, 52-tetracosaazanonadecacyclo$[41.11.1.1^{7,19}.1^{10,52}.1^{16,28}.1^{25,37}.1^{34,46}.0^{3,53}.$ $0^{5,9}.0^{8,12}.0^{14,18}.0^{17,21}.0^{23,27}.0^{26,30}.0^{32,36}.0^{35,39}.0^{41,45}.0^{44,48}.0^{50,54}]$hexacontane-2, 6, 11, 15, 20, 24, 29, 33, 38, 42, 47, 51-dodecone.

3

NMR spectroscopy [6, 7]. Although some simple aliphatics do bind [7], most work has been done with alkylammonium ions, which show exceptional ligand-receptor affinity. To a certain extent this has been a matter of practicability, since cucurbituril only dissolves appreciably in aqueous acidic solution (HCO_2H–D_2O has been adopted as the "standard" solvent). It seems that protonated-amine cations coordinate to the occuli of cucurbituril, due to favorable ion-dipole attractions cumulatively involving the urea carbonyls. Should a suitably-sized hydrocarbon substituent be attached also to an ammonium ion that is engaged in such fashion, it may enter the internal cavity, displacing and freeing solvent molecules and making a hydrophobic contribution to complexation. This coupling of recognition factors lends great specificity to the capture of alkylammonium ions. An illustrative sampling of the binding behavior exhibited by cucurbituril ensues.

2.2 Structure and Selectivity

2.2.1 Evidence for Internal Binding [7]

The ^1H NMR spectrum of 1,5-diaminopentane in HCO_2H–D_2O solution exhibits two CH resonance multiplets at $\delta\,3.17$ and 1.77. These correspond respectively to the methylene units adjacent to ammonium ions (4 H) and to the remaining, or central, methylene groups (6 H). Addition of small portions of cucurbituril to such a solution leads to a progressive replacement of these signals by two new multiplets, at $\delta\,2.73$ and 0.77. Eventually, a stoichiometric adduct (1:1 mole ratio) may be obtained. When an excess of the diamine ligand is present, both sets of NMR signals are seen, indicating that exchange between bound and free pentanediammonium ions is slow on the NMR time scale. This is a common feature of cucurbituril complexes, and greatly facilitates relative affinity measurements for guests by NMR spectral integration, as described subsequently. The upfield chemical shifts upon complexation are characteristic, and indicate that the interior of cucurbituril comprises a proton-shielding region relative to the acidic aqueous medium employed for solvating the host species.

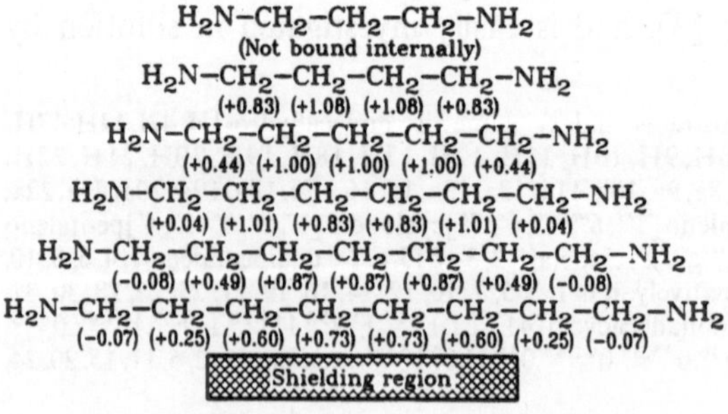

$$H_2N-CH_2-CH_2-CH_2-NH_2$$
(Not bound internally)

$$H_2N-CH_2-CH_2-CH_2-CH_2-NH_2$$
(+0.83) (+1.08) (+1.08) (+0.83)

$$H_2N-CH_2-CH_2-CH_2-CH_2-CH_2-NH_2$$
(+0.44) (+1.00) (+1.00) (+1.00) (+0.44)

$$H_2N-CH_2-CH_2-CH_2-CH_2-CH_2-CH_2-NH_2$$
(+0.04) (+1.01) (+0.83) (+0.83) (+1.01) (+0.04)

$$H_2N-CH_2-CH_2-CH_2-CH_2-CH_2-CH_2-CH_2-NH_2$$
(−0.08) (+0.49) (+0.87) (+0.87) (+0.87) (+0.49) (−0.08)

$$H_2N-CH_2-CH_2-CH_2-CH_2-CH_2-CH_2-CH_2-CH_2-NH_2$$
(−0.07) (+0.25) (+0.60) (+0.73) (+0.73) (+0.60) (+0.25) (−0.07)

Shielding region

Fig. 2. ^1H NMR induced shifts (ppm) of methylene groups of alkanediammonium ions upon complexation with cucurbituril (D_2O–HCO_2H solution); from [7] with permission

This NMR technique has been applied to a series of alkanediammonium ions, and the results (induced shifts of proton resonances) are summarized in Fig. 2. It may be seen that the shielding region extends for approximately 4.5 methylene units, or 6 Å, which coincidentally is the interatomic distance axially spanning the cavity of cucurbituril. Similar induced chemical shift effects are found in ^{13}C NMR spectra, and UV spectral perturbations are noted upon encapsulation of certain aromatic-ring bearing ammonium ions (particularly 4-methylbenzylamine). Conclusive evidence for internal complexation with cucurbituril has been secured by crystallography [3].

2.2.2 Competitive Complexation

In an effort to define practically the dimensions of the host cavity, we have surveyed the complexing capability of cucurbituril toward a variety of potential guests. Formation constants for over 60 substituted ammonium ion ligands of this receptor have been ascertained [6, 7]. This has been achieved mainly by competitive NMR experiments in which an excess of two different RNH_3^+ cations was allowed to compete for a limited amount of cucurbituril in the standard solvent. The composition of the resulting mixture was then determined by accurate NMR integration. This method was supplemented by a similar quantitative UV technique for guests showing a spectral perturbation on complexation. By use of such methods, with appropriate relays between ammonium ions for which cucurbituril has very dissimilar captivation potential, an extensive compilation of relative affinities has been made available [7]. An abbreviated list of comparative binding efficiencies is provided in Table 1.

The data cumulatively lead to the following general conclusions. Within homologous series of n-alkaneamines, the pattern of affinity for cucurbituril is one of gradual enhancement as chain length is increased until a maximum is

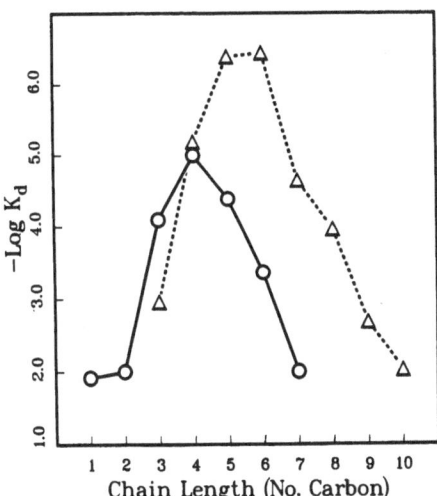

Fig. 3. Dependence of strength of binding of cucurbituril upon chain length for n-alkylammonium ions (O–O) and n-alkanediammonium ions (△–△). Vertical axis is proportional to free energy of binding (log K_d); from [7] with permission

Table 1. Affinity data for ligand-receptor complexes of cucurbituril [7]

No.	Ammonium ion ligand	$K_f(rel)^a$	No.	Ammonium ion ligand	$K_f(rel)^a$
1.	NH_3	0.25		diammonium ions:	
			25.	$NH_2(CH_2)_3NH_2$	2.8
	simple alkyl substituents:		26.	$NH_2(CH_2)_4NH_2$	480
2.	CH_3NH_2	0.25	27.	$NH_2(CH_2)_5NH_2$	7600
3.	$CH_3CH_2NH_2$	0.3	28.	$NH_2(CH_2)_6NH_2$	8600
4.	$CH_3(CH_2)_2NH_2$	37.6	29.	$NH_2(CH_2)_7NH_2$	135
5.	$CH_3(CH_2)_3NH_2$	307	30.	$NH_2(CH_2)_8NH_2$	28
6.	$CH_3(CH_2)_4NH_2$	74	31.	$NH_2(CH_2)_9NH_2$	1.5
7.	$CH_3(CH_2)_5NH_2$	7.0	32.	$NH_2(CH_2)_{10}NH_2$	0.3
8.	$CH_3(CH_2)_6NH_2$	0.3			
9.	$(CH_3)_2CHCH_2NH_2$	67		arene-containing substituents:	
10.	$(CH_3)_2CH(CH_2)_2NH_2$	109	33.	$C_6H_5CH_2NH_2$	0.8
11.	$(CH_3)_2CH(CH_2)_3NH_2$	13	34.	$p\text{-}CH_3C_6H_4CH_2NH_2$	1.0
12.	$CH_3CH_2CH(CH_3)CH_2NH_2$	6.0	35.	$(2\text{-}C_4H_3S)CH_2NH_2$	710
13.	$CH_3CH_2CH(CH_3)(CH_2)_2NH_2$	3.5	36.	$(2\text{-}C_4H_3O)CH_2NH_2$	350
	cyclo-alkyl substituents:			miscellaneous:	
14.	$cyclo\text{-}(CH_2)_2CHCH_2NH_2$	45	37.	$(CH_3)_3C(CH_2)_2NH_2$	< 0.1
15.	$cyclo\text{-}(CH_2)_3CHCH_2NH_2$	1130	38.	$CH_3(CH_2)_3NHCH_3$	340
16.	$cyclo\text{-}(CH_2)_4CHCH_2NH_2$	1040	39.	$CH_3(CH_2)_3N(CH_3)_2$	2.3
17.	$cyclo\text{-}(CH_2)_2CHNH_2$	1.2	40.	$NH_2(CH_2)_6OH$	3.6
18.	$cyclo\text{-}(CH_2)_3CHNH_2$	9.2	41.	$HC \equiv CCH_2NH_2$	4.8
19.	$cyclo\text{-}(CH_2)_4CHNH_2$	19.5	42.	$N_3(CH_2)_2NH_2$	1.2
			43.	$NH_2(CH_2)_4NH(CH_2)_3NH_2$	4200
	thioether-containing substituents:		44.	$NH_2(CH_2)_3NH(CH_2)_4NH(CH_2)_3NH_2$	
20.	$CH_3S(CH_2)_2NH_2$	52			40000
21.	$CH_3CH_2S(CH_2)_2NH_2$	105			
22.	$CH_3S(CH_2)_3NH_2$	27			
23.	$CH_3CH_2S(CH_2)_3NH_2$	2.3			
24.	$cyclo\text{-}(CH_2S)_2CHCH_2NH_2$	1810			

^{a} Formation constant relative to no. 34, for which absolute value of $K_d = 3.1$ mM in aqueous formic acid, 40 °C.

reached, followed by diminished affinity upon further extension. This is conveniently seen graphically in Fig. 3. For substrates $H(CH_2)_nNH_3^+$, n-butylamine forms the most stable complex (n = 4) and the order of affinities for cucurbituril follows the trend n = 1 < 2 < 3 < 4 > 5 > 6 > 7. As subsequently amplified, the explanation is that the butyl substituent optimally fills the cavity, and longer chains are obliged to protrude into the second occulus of cucurbituril, interfering there with solvation of the carbonyl dipoles. A similar pattern is observed with the n-alkanediamines, for which a hydrocarbon chain length of 5 or 6 is best (Fig. 3, dashed line). In this series guests that are longer apparently encounter difficulty in simultaneously coordinating both ammonium ions to occuli of cucurbituril. A rapid shuttling of the receptor along the hydrocarbon chain between nitrogens is suggested by a "smearing out" of the NMR shielding zone as may be seen particularly for octanediamine from the data in Fig. 2.

Although the nominal opening within the occuli of cucurbituril is only 4 Å in diameter, the internal cavity calculates to be somewhat larger than 5 Å across, suggesting that groups of greater dimensions than a polymethylene chain could be successfully incorporated. This is the case; complexation has been observed with isobutyl- and isopentylamines, as well as with cyclopropane-, cyclobutane-, and cyclopentane-substituted amines. Such examples indicate that the cavity can accommodate a branching methyl group or an aliphatic ring size up to five. On the other hand, cyclohexanemethylamine and n-alkaneamine chains bearing an ethyl substituent or more than one branching methyl group (e.g. $(CH_3CH_2)_2CHCH_2NH_3^+$, $(CH_3)_3CCH_2NH_3^+$, $(CH_3)_2CHCH(CH_3)CH_2NH_3^+$, etc. . .) all showed no perturbation of NMR peak positions in the presence of cucurbituril. Apparently they are too bulky to form an internal complex. The benzene ring, which has van der Waals dimensions (6-Å diameter × 4-Å thick) larger than the estimated internal cavity, does incorporate satisfactorily when attached to a suitable cation. This constitutes a special case to be considered subsequently. A cross-sectional representation of the cyclopentanemethylammonium ion complex of cucurbituril is presented in Fig. 4, demonstrating optimal filling of the internal cavity by the ligand, and with the cationic group in registry (H-bonding) with the receptor carbonyl groups. In the latter regard, cyclopentylammonium ion (lacking the CH_2 between ring and cation) binds

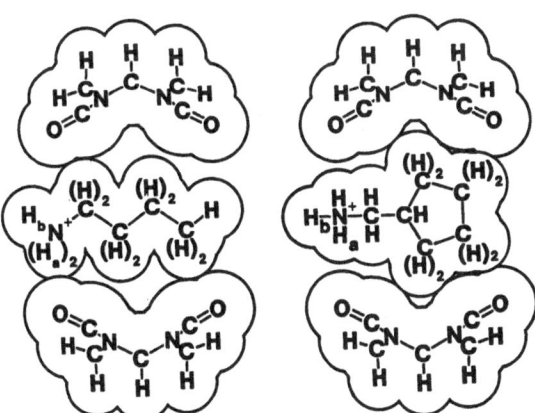

Fig. 4. Conjectured cross-sectional representation of complexes of cucurbituril plus n-butylammonium ion, as well as with cyclopentanemethylammonium ion. Outlines are drawn to van der Waals radii (maximum projection for all atoms upon axial rotation of the receptor, crystallographically determined interatomic distances for cucurbituril). Two N–H···O=C hydrogen bonds may form, but the third H–N$^+$ projects incorrectly for coordination to the receptor; from [7] with permission

53-fold less readily than the species shown, and pyrrolidinium (cation within the ring) apparently binds not at all.

Such considerations suggest that proper alignment of bound ammonium ions with the host carbonyl dipoles is critical. Since there are six of the latter surrounding each occulus of cucurbituril, it would seem that a tripodal H-bonding scheme is possible, employing the oxygen atoms of alternate carbonyls as H-bond receptors from the RNH_3^+ ion. Indeed, this should be the case for very small guests (e.g. $CH_3NH_3^+$). However, closer examination of structural models such as Fig. 4 suggests that this may not be valid for guests which fill the cavity of cucurbituril. The difficulty is that such a symmetrical H-bond network requires a C–N bond within the guest which is *colinear* with the central axis of the host. For the *n*-butyl group in a staggered conformation, the C–N bond must be tilted *off*-axis in order to avoid unacceptable van der Waals contacts between the rest of the guest and the interior of the cavity of cucurbituril (cf. Fig. 4). As a consequence only *two* of the protons on nitrogen (H_a in Fig. 4) may contact carbonyl oxygens, and the third (H_b) projects away from the occulus. This somewhat esoteric point is susceptible to experimental test. Upon examining the ligation of *n*-butylammonium ions to cucurbituril, one finds that n-$C_4H_9NH_3^+$ and n-$C_4H_9NH_2^+CH_3$ bind equally well, whereas n-C_4H_9-$NH^+(CH_3)_2$ binds >1000-fold more poorly. This reveals that *one* of the H–N linkages of *n*-butylamine may be replaced by an alkyl residue without detriment to binding (i.e. CH_3 may occupy the H_b site), but substitution of two H–N bonds by methyls strongly perturbs binding (by invading the H_a site). The same trend is noted for the complexation of hexanediamines [i.e. $NH_3^+(CH_2)_6NH_3^+ \approx CH_3NH_2^+(CH_2)_6NH_2^+CH_3 \gg (CH_3)_2NH^+(CH_2)_6NH^+$-$(CH_3)_2$].

However, we believe that the pair of hydrogen bonds so disclosed at each occulus is only indirectly of consequence energywise. This conclusion follows from an observation that whereas $NH_3^+(CH_2)_6NH_3^+$ binds 1000-fold more tightly to cucurbituril than does $H(CH_2)_6NH_3^+$, complexation of HO-$(CH_2)_6NH_3^+$ is actually weaker than for the *n*-hexylammonium ion. Evidently the hydroxyl group contributes nothing to stabilization of the adduct, in spite of the fact that an alcohol is also a good H-bonding substituent. By way of explanation, the important consideration would appear to be the *difference* in stabilization of free and complexed guests. While the alcohol as well as the ammonium ions may be H-bonded in the complex, in the absence of the receptor they would also be fully H-bonded within the polar aqueous solvent employed. The consequential feature of ammonium ions is that they are *charged*. The occuli of cucurbituril represent a region of negative charge accumulation, since the anionic ends of the dipoles of six urea carbonyl groups are focused there. Hence, it is our understanding that the high specificity for ammonium ions is largely an electrostatic *ion-dipole attraction*. H-bonding may occur, but this is incidental; it merely allows the closest contact between cation and dipole.

Binding of arylamines constitutes the most sensitive test of the internal capacity of cucurbituril. A complex is formed with p-$CH_3C_6H_4CH_2NH_3^+$, but not with its o- and m-isomers: the methyl group must be able to orient into the second occulus. Benzylammonium ion itself binds relatively weakly, but thiophenemethylammonium ion, which has dimensions closer to that of the internal diameter of the host, has a 700-fold higher affinity. This suggests that a benzene ring exceeds the strain-free binding capacity of the receptor. The latter idea finds confirmation in the crystal structure of the p-NH$_3^+$CH$_2$C$_6$H$_4$-CH$_2$NH$_3^+$ complex of cucurbituril [3]. The cage structure of the host is clearly distorted into an ellipsoid shape in this adduct, with more than a 0.4-Å decrease in diameter (compared with uncomplexed cucurbituril) in a direction perpendicular to the guest benzene ring and with a compensating increase within the plane of the aromatic ring. Hence, the relatively low affinities of benzenoid guests reflects a balanced compensation between favorable noncovalent binding forces and a stress-strain relationship involving host and guest. If the difference between the stability of the complexes of C$_6$H$_5$CH$_2$NH$_3^+$ and (2-C$_4$H$_3$S)CH$_2$-NH$_3^+$ is taken as a measure of misfit, then there is a ΔG of 4.2 kcal/mol which should be partitioned as 2.1 kcal/mol distortion energy in cucurbituril (strain) and 2.1 kcal/mol of compression energy (stress) within a bound benzene. In any event, a practical limit to the inclusion capacity of cucurbituril corresponds to the volume of a benzene ring.

2.2.3 A Molecular Switch

The availability of quantitative guest-affinity data as previously discussed allows the engineering of molecular devices based on cucurbituril. An example is the construction of the "molecular switch" (defined as a ligand-receptor system which has the capacity to exist in more than one state, contingent upon some controlling element) that is diagrammed in Fig. 5. Triamine ligand C$_6$H$_5$NH-(CH$_2$)$_6$NH(CH$_2$)$_4$NH$_2$ was specifically designed and prepared so as to be capable of binding in two distinct ways to cucurbituril [8]. The important feature of the guest is that the pair of nitrogen atoms not connected directly to a benzene ring ought to be 10^6-fold more basic than the one which is. The other piece of relevant information is that hexanediammonium ion binds 100-fold more tightly than butanediammonium ion with cucurbituril (Fig. 3). The pK_a of the anilinium nitrogen of the ligand, which has a value of 4.7 in aqueous solution, increases to a value of 6.7 in the presence of a stoichiometric amount of cucurbituril (as measured by spectrophotometric titration). Such behavior was quite predictable, and its cause was revealed by NMR spectroscopy of the complex as a function of pH. In acidic solution cucurbituril adheres to the ligand by spanning the hexanediamine portion, but at a pH of >6.7 it coordinates to the butanediamine end as in Fig. 5 (bottom). So long as the aniline nitrogen remains protonated, binding is favored across the six-carbon site because of its greater complementarity to the interior dimensions of the receptor, with both occuli being occupied with secondary ammonium ions, Fig. 5 (top). Upon deprotonation of the aniline nitrogen, the receptor translocates to the four-carbon site of the ligand. While binding with a butanediammonium ion may be

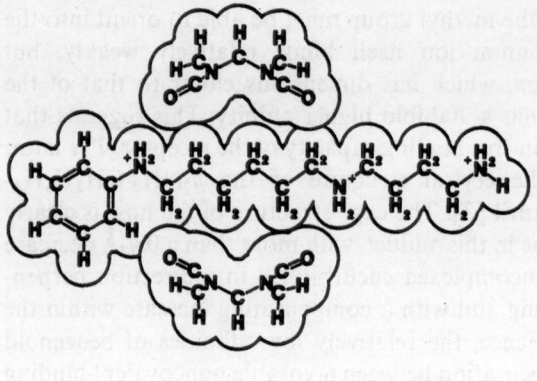

pH>6.7 ‖ pH<6.7

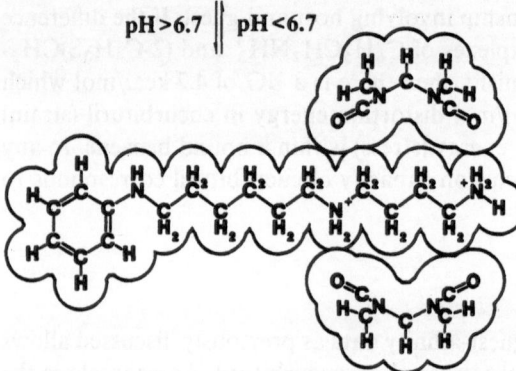

Fig. 5. Conjectured cross-sectional representation of complex for cucurbituril plus $C_6H_5NH(CH_2)_6NH-(CH_2)_4NH_2$ in acidic and in alkaline solution. Outlines are drawn as for Fig. 4. In all probability a slight buckling of the hexyl chain actually occurs in the upper situation, better filling the cavity and bringing the ammonium ions into improved alignment with the carbonyl dipoles of the receptor, from [8] with permission

intrinsically less favorable, it is superior to that of an *n*-hexyl(*mono*)ammonium ion. The evidence indicates that coordination at the subsidiary site is 100-fold less stable thermodynamically, since a 2 pH-unit bias is necessary to drive the receptor to this position from its favored location. In this "switch" a hydronium ion is functioning as a control element, inducing translocation between two ligation states, so as to maintain optimal ion-dipole interactions.

2.3 Quantitative Structure-Activity Correlations

The availability of formation constants for a large number of structurally diverse alkylammonium ions also encourages attempts to identify quantitatively those chemical features of a resident ligand species that are responsible for stability and selectivity within complexes of cucurbituril. In order to secure interpretable results, a subset of data was compiled, restricted to mono-substituted ammonium ions carrying exclusively alkyl or thioether residues [9]. Only included were moieties for which internal complexation with cucurbituril was established (NMR evidence) and for which substrate-specific overcrowding (such as with the arenes) was unlikely. A short list consisting of the first 24 guests in Table 1 was found to meet these criteria. The intention was to evaluate systematically those factors governing the *hydrophobic* component of molecular recognition in this system. All substrates were monoammonium ions, so that the ion-dipole attraction with the urea carbonyls should be a constant feature, cancelling out upon data reduction.

Initial attempts to correlate the affinity of ligands toward cucurbituril using independently estimable parameters such as van der Waals molecular surface area or molecular volume of the guests were relatively unsuccessful. The difficulty apparently is that the interior of the receptor has a definite shape and distribution of polarity, so that complementarity between cucurbituril and its ligand depends more subtly upon structure of the bound entity. Consequently we opted for an empirical treatment of our data, which would yield an indication of how particular regions of the interior of cucurbituril interact with ligands.

A major complication in interpreting the disparate affinities of various ligands for cucurbituril is that the experiment K_f values that we have obtained represent differences between stabilization energy of the complexes and the solvation energy of the individual components prior to association. Consequently, a large differential between the strength of complexation of any two alkylammonium ions with cucurbituril might actually reflect major differences in their solvation in the absence of receptor, rather than any very great difference in the absolute stabilities of the complexes themselves. This seems particularly so, when it is realized that ligands are effectively sequestered from solvent when ensconced within cucurbituril (cf. Fig. 4). Therefore, a systematic treatment of comparative affinities for cucurbituril must first account for solute-solvent effects in the free (unbound) ligands.

A resolution for this problem has been identified by Hine [10]. In principle, the reference state for the uncomplexed ligand should be the dilute vapor phase, rather than in solution, so that all intermolecular contacts are negated in the dissociated state. In such a case the relative affinities of ligands for cucurbituril would solely reflect stability of the complexes. Although such a measurement is not directly feasible in our case, data exists that allows computation of an energy term for solvation of each of the employed ligands. From empirical source data consisting of aqueous solubilities and vapor pressures of model organic substances, Hine has provided a group-contribution scheme yielding the intrinsic hydrophobicity of various fragments of an organic structure (CH_3-, $-CH_2-$, $-CH<$, $-S-$, etc..., in various molecular environments). Using these parameters one can make a *solvation correction* for each of the ligands to be correlated (i.e. group-transfer equivalents from aqueous solution to the vapor phase). When this is done a reduced data set is produced, in terms of free energies of binding, which is likely to reflect primarily the fit of substrates within cucurbituril [9].

In order to process the emended data in as unprejudiced a manner as possible, we adopted as our fundamental variable for correlating structure with affinity the intramolecular *distance* of an alkyl fragment of a bound ligand from its ammonium ion. The rationale for this approach is that the binding site for the

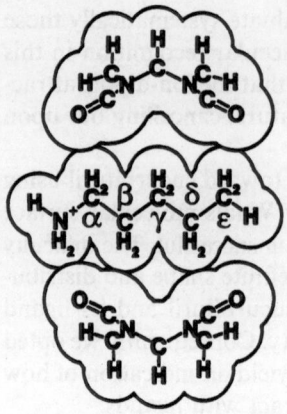

Fig. 6. Conjectured cross-sectional representation of complex for cucurbituril plus *n*-pentylammonium ion. Outlines are drawn as for Fig. 4. Methylene groups are labelled according to their relationship to the ammonium residue (α–ε); from [9] with permission

RNH_3^+ moiety is established and is presumably invariant (for the most part); i.e. the ammonium ion coordinates with the carbonyl dipoles of cucurbituril as depicted in Fig. 6. We then inquire as to the *contributions* of hydrocarbon (or thioether) fragments in the $\alpha, \beta, \gamma, \delta$, and ε positions, as also noted in Fig. 6, to the stability of the ligand-receptor complexes. So as to do this most efficiently, we simply count hydrogen atoms at each of these positions. Since the hydrogens represent regions of contact with the receptor, they provide an elementary index of potential interaction with the interior of cucurbituril. No distinction is made between primary (CH_3), secondary (CH_2), or tertiary (CH) hydrogens in our analysis, i.e. a methylene group is assumed to be twice as consequential as a methine moiety. In the case of branched or cyclic alkyl substituents, the number for a particular type of hydrogen is incremented accordingly.

The purpose of this breakdown is to perform a *regression analysis* upon corrected binding energies, and thereby to identify how the *location* of each individual ligand-CH fragment influences stability of the receptor complexes. In order to get meaningful results from such an approach, it is necessary to restrict the number of disposable parameters to a minimum. Preliminary multiparameter fitting indicated that certain individual CH-fragment contributions could be combined for purposes of simplification. It appeared that hydrogens in the γ and δ positions (Fig. 6) were the major hydrocarbon contributors to stabilization of the complexes and that their contributions were comparable. Accordingly, these parameters were united. Conversely, hydrogens in the α or the ε and $\varepsilon +$ positions (i.e. immediately next to, or more than four atoms removed from the nitrogen, as in *n*-pentyl- through *n*-heptylamine) made an unfavorable contribution to ligand-receptor interactions. These too were combined into a single parameter. The influence of hydrogens in the β position (Fig. 6) appeared to be nearly neutral, and these were consequently ignored in the final fit. Sulfur atoms (thioether linkages in the γ or δ positions) made a strong positive contribution, and they were assigned to a separate parameter. Ultimately, an adequate correlation of binding energies for all 24 substrates was secured with just these three adjustable parameters (plus a catchall fourth term common to each, incorporating the ammonium residue and other essentially constant but unevaluable factors, such as desolvation of the receptor upon cation complexation).

The upshot of this analysis is that a methylene group within the center of the cavity inside cucurbituril (i.e. in the γ or δ positions, Fig. 6) provides beneficial noncovalent binding amounting to ca. -0.76 kcal/mol (ΔG), according to the data treatment indicated. It has been independently estimated that transformation of a $-CH_2-$ moiety from the vapor phase to a hydrocarbon liquid is exergonic by -0.82 kcal/mol (from heats of vaporization). Hence, the central interior of cucurbituril appears typically "hydrocarbon-like" (since our ΔG is also referenced to the vapor state). Placing a sulfur atom in that same location appears to result in an even greater stabilization ($\Delta G = -5.3$ kcal/mol per S). This finding matches theoretical expectations based upon the relative polarizabilities of $-CH_2-$ and $-S-$, which suggest the availability of greater dispersion forces at the interface in the latter case. However, the practical consequences of this phenomenon as regards hydrophobic binding are considerably less, for the $-S-$ group has a compensating enhanced *hydrophilicity* according to Hine's data, a point which has not always been appreciated [9]. On balance, the $-CH_2-$ and $-S-$ groups are surrogates for one another, as can be seen by comparing K_f values for isostructural ligands in Table 1.

Perhaps the most telling feature of the factor analysis is a ΔG of $+0.76$ kcal/mole associated with placing a methylene group within the vicinity of the oxygen atoms of cucurbituril (i.e. in the α or ε positions). Forcing a $-CH_2-$ to reside near the occuli of the receptor apparently has a *destabilizing* effect, which coincidentally is of same magnitude but of the opposite direction as placing it in the center of the cavity. The polarized regions surrounding the carbonyl groups (adjacent to and opposite from the ammonium ion binding site in Fig. 6) reject hydrocarbons. Apparently it is the concurrent interactions of these neighboring regions of the receptor that are responsible for the exceptional selectivities that cucurbituril exhibits toward hydrocarbon-containing guests. Displacing water from the polarized region of the carbonyls, as required of lengthier alkyl chains which protrude into the second occulus, provides endergonicity comparable to the exergonicity of filling the cavity with methylene groups. It is the close juxtaposition of hydrophobic and hydrophilic regions within cucurbituril that *doubles* the selectivity that is ordinarily obtainable in transferring hydrocarbons from aqueous to lipid-like environment. It seems highly likely that biological receptors should be able to take advantage of this phenomenon. Proteins are replete with the appropriate functionality (hydrocarbon side chains plus carboxamide dipoles). In this respect, cucurbituril is a uniquely informative biochemical model system.

In principle, a thermodynamic value for the ion-dipole interaction could be acquired as well by the incorporation of several alkanediamines into our data set in the foregoing analysis. However, a second ammonium ion has in fact a disproportionate effect, and it cannot be meaningfully placed on the same scale as the hydrocarbon fragment increments previously considered (at least not with reference to the vapor phase). For what it may be worth, 1,5-pentanediamine binds more tightly than *n*-pentylamine by $-2.8 \, \text{kcal/mol}$ (unemended ΔG). This may be a reasonable estimate for the attactive contribution to binding by six carbonyl dipoles interacting with one ammonium ion (over and above the stabilization provided by aqueous formic acid).

2.4 Dynamics of Binding

So far, the studies of cucurbituril described have been *thermodynamic* investigations, in which factors contributing to the overall stability of molecular complexes have been explored. While bounteous, these only partly address the question of receptor specificity. For example, in biological systems the *kinetics* of noncovalent interactions, such as between enzymes and substrates, may be of greater consequence. Clearly, the *dynamics* of molecular recognition deserve additional attention. Cucurbituril provides diverse opportunities in this area [11].

2.4.1 Guest Displacement Mechanism

As previously indicated, for many cucurbituril ligands the exchange between the interior cavity and external solvent is slow on the NMR time scale (separate signals for bound species and for excess ligand). In the course of collecting affinity data, it was noted that in a number of instances equilibrium was only slowly attained when a second ligand was added to a preexisting solution of cucurbituril complexed with an initial ligand. The change in NMR signal intensity was observed to follow an exponential decay. This makes feasible systematic investigation of the kinetics of ligation.

Because there are two opposed occuli providing entrances to the cavity within cucurbituril, a concerted substitution is imaginable whereby a preexisting ligand is pushed out by its replacement. Were this *associative* process to take place, the reaction velocity should accelerate proportionately upon increase in the concentration of displacing agent (second-order reaction). However, the experimental results were otherwise. For example, when $(CH_3)_2CHCH_2$-$CH_2NH_3^+$ within cucurbituril is replaced by a molar equivalent of $(2\text{-}C_4H_3O)$-$CH_2NH_3^+$, a (pseudo)first-order rate constant of $0.37 \times 10^{-3} \, \text{s}^{-1}$ was measured; a 10-fold increase in the concentration of the second species actually yielded a small rate decrease, to $k = 0.24 \times 10^{-3} \, \text{s}^{-1}$. Similar results were obtained upon variation of concentration with a number of other mono- and diammonium ion ligand pairs. It must be concluded that the substitution is a *dissociative* process: in a slow step the initial guest escapes from the cavity (or is displaced by solvent), with a subsequent competition between ammonium ion species for the free cage molecule.

Another detail of ligand dissociation from cucurbituril commands interest, particularly in the case of alkanediamines. Obviously one of the nitrogens of a bound alkanediammonium ion must be drawn through the core of the molecule. With the nitrogen protonated, desolvation of the charged moiety as it passes into an apparently "hydrocarbon-like" environment should contribute substantially to the activation energy of the process, but transient deprotonation of one nitrogen could obviate the problem. This should show up as a pH dependence for the rate of ligand exchange. Since rate measurements are obtained in buffered medium (HCO_2H/HCO_2^-), this uncertainty ought to be resolvable. Upon a pH increase of one unit, there would necessarily be a 10-fold increase in the small equilibrium amount of (unprotonated) alkylamine complexed in acidic solution. However, the experimental observation is for no significant velocity dependence upon pH, for either an alkanediammonium ion or for a simple alkyl(mono)ammonium ion in buffered aqueous formic acid solution (pH values of 1–4). This suggests that deprotonation is not required for ligand dissociation from cucurbituril to proceed.

2.4.2 Rate Dependence Upon Ligand Size

Because of the constriction provided by the occuli of cucurbituril (diameter 4 Å), ligands which entirely fill the cavity (cf. Fig. 4) might be expected to exchange more slowly. A systematic investigation of this phenomenon is summarized in Table 2. The relevant numbers are the values of k_{in}, the second-order rate constant for bimolecular formation of the complexes [11]. Although these can be measured directly by UV spectrophotometry for the slow incorporation of

Table 2. Ligand-receptor kinetic data for cucurbituril

No.	Ligand[a]	$10^5 K_d$, M	$10^5 k_{out}$, s^{-1}	k_{in}, s^{-1} M^{-1}	T_c, °C
1.	p-CH$_3$C$_6$H$_4$CH$_2$NH$_2$	310	850	2.7	
2.	(CH$_3$)$_2$CH(CH$_2$)$_2$NH$_2$	2.8	37	13.3	
3.	(CH$_3$)$_2$CH(CH$_2$)$_3$NH$_2$	24	120	5.0	
4.	(2-C$_4$H$_3$S)$_2$CH$_2$NH$_2$	0.43	9.3	21.4	
5.	$cyclo$-(CH$_2$S)CHCH$_2$NH$_2$	0.17	1.6	9.2	
6.	$cyclo$-(CH$_2$)$_4$CHCH$_2$NH$_2$	0.30	1.6	5.5	
7.	$cyclo$-(CH$_2$)$_3$CHCH$_2$NH$_2$	0.27	1600	5900	
8.	$cyclo$-(CH$_2$)$_2$CHCH$_2$NH$_2$	6.8	$>10^{7b}$	$\geq 10^6$	<40
9.	CH$_3$(CH$_2$)$_3$NH$_2$	1.0			74
10.	CH$_3$(CH$_2$)$_4$NH$_2$	4.2			89
11.	CH$_3$(CH$_2$)$_5$NH$_2$	44			89
12.	CH$_3$(CH$_2$)$_6$NH$_2$	990			45

[a] In HCO_2H/D_2O solution, 40 °C (data extrapolation from lower temperature for no. 1). [b] Estimated from NMR coalescence behavior [11].

15

arylammonium ions (entry no. 1), more generally they are determined from the ratio k_{out}/K_d, the rate constant for dissociation (measured by NMR as previously described) divided by the equilibrium dissociation constant (measured by competition between ligands). The directly determined rate constant k_{out} is less suitable for evaluating the transition state leading to complexation, since its values should also reflect the differential stabilities of the complexes themselves. The general pattern is that branching within the alkyl portion of the substrate uniformly retards insertion of a ligand into cucurbituril.

A straightforward connection exists between the thickness of the ligand and its rate of complex formation with cucurbituril. The trend is succinctly illustrated by the cycloalkanemethylammonium ions, entries no. 6–8. The five-membered carbocycle, with a calculated maximum van der Waals diameter of 5.7 Å (i.e. across the 2,5-positions of the ring), forms its adduct most slowly. The smaller four-membered ring binds more speedily by three orders of magnitude. The three-membered ring is even more rapid, as are the n-alkylammonium ions, diameter 4 Å. In the latter cases displacement rates are not measurable by the technique indicated. However, *coalescence* behavior is seen in the NMR spectra upon variable temperature examination of cucurbituril complexes in the presence of an excess of the ligand. This indicates rapid exchange, which apparently does not depend greatly upon the length of the n-alkyl chain (T_c = coalescence temperature). Unbranched hydrocarbon chains, with a diameter matching that of an occulus, clearly may slip in without impediment, but the larger-dimensioned rings and isoalkyl hydrocarbon moieties are obliged to force their way into cucurbituril, with probable distortion of the opening in the receptor.

2.4.3 Absence of Dynamic Cohesion

The molecular motions of guests ensconced within cucurbituril are of interest. Information on the relative tumbling rates in solution of receptor and ligand is accessible from NMR spectroscopy [11]. The predominant relaxation process for ^{13}C nuclei within CH_2 and CH groups arises from dipolar interactions with directly attached hydrogen nuclei. For a ^{13}C nucleus the measured spin-lattice relaxation time T_1 yields a correlation time t_c for its motions according to the equation $t_c = (2.8 \times 10^{-11})r^6/nT_1$, in which n is the number of hydrogens attached to a particular carbon at a distance r. Suitable carbon atoms are present in both cucurbituril and many of its ligands. Hence, NMR relaxation rate measurements on the complexes allow an examination of whether stability of a complex (as measured by the quantitative affinity data) is reflected in the relative freedom of guest to move about within the host.

Typically, t_c values for the receptor carbons fall in the range of 200–300 ps, which is a reasonable value for a molecule of its size, with guest complexation not providing any perturbation, as might have been expected. For more rapidly tumbling alkylammonium ions of the type incorporated, t_c values for CH_2 and CH groups are typically 4–10 ps in the uncomplexed state. For a number of ligands this value was found to be raised upon complexation with cucurbituril, but only to values amounting to $20(\pm 6)\%$ of that of the receptor. This appeared to be true regardless of the structure of the bound species. A diamine [$NH_3^+(CH_2)_5NH_3^+$] behaved the same as an n-alkylamine [$CH_3(CH_2)_3NH_3^+$], and ligands known to fill the cavity optimally [$cyclo$-$(CH_2)_3CHCH_2NH_3^+$, $(2\text{-}C_4H_3S)CH_2NH_3^+$] exhibited the same tumbling frequency. Even a bound benzene-containing guest ($C_6H_5CH_2NH_3^+$), which previous evidence has suggested is squeezed by the host when incorporated (cf. Sect. 2.2.2), did not show a correlation time approaching that of the receptor. (All test ligands were known not to exchange rapidly with the exterior solvent.)

This indicates a lack of *dynamic cohesion* within the adducts; i.e. the substrate has considerable freedom for reorientation within the receptor. The apparent reason for an absence of mechanical coupling is the nearly cylindrical symmetry of cucurbituril, which allows the guest an axis of rotational freedom when held within the cavity. Hence, the bound substrates show only a moderate increase in t_c relative to that exhibited in solution. No relationship exists between t_c values and the thermodynamic stability of the complexes as gauged by K_f (or K_d, cf. Tables 1 and 2). It must be concluded that the interior of cucurbituril is notably "nonsticky". This reinforces previous conclusions that the thermodynamic affinity within adducts is chiefly governed by hydrophobic interactions affecting the solvated hydrocarbon components, plus electrostatic ion-dipole attractions between the carbonyls of the receptor and the ammonium cation of the ligands.

3 Catalysis

A major goal of investigators in supramolecular chemistry is enzyme mimesis. Understanding biocatalysis represents perhaps the most severe challenge of mechanistic organic chemistry. Certainly a major component of enzyme efficiency is the negation of entropic constraints through proper alignment of reacting substrate functionalities. However, additional catalytic benefit is in theory available to ligand-receptor systems which have potentially an *excess* of binding energy. The Pauling principle of catalysis states that complementarity between an enzyme and the *transition state* for its conducted reaction ought to be greater than that between enzyme and the reactants [12]. In such circumstances an additional promotion of the chemical transformation is attained, resulting in rates beyond that attributable to orientation considerations alone. The latter concept has assumed the status of dogma in theoretical enzymology [13], yet experimental verification has lagged. Consequently, acquisition of evidence in support of this principle ought to be a critical objective.

3.1 Azide-Alkyne 1,3-Dipolar Cycloaddition

Based on accumulated knowledge of the molecular recognition characteristics of cucurbituril, the following reaction was selected for a systematic investigation. Alkynes undergo 1,3-dipolar cycloaddition with alkyl azides, yielding substituted triazoles, Eq. (2). In the particular case of propargylamine plus azidoethylamine, the reaction proceeds slowly ($k_0 = 1.16 \times 10^{-6}\,M^{-1}s^{-1}$) in the standard solvent, yielding a pair of regioisomeric adducts in equal amount [14, 15]. Previous investigations of this type of transformation have shown it to be a typical concerted pericyclic reaction. No evidence for the occurrence of any intermediate species has emerged.

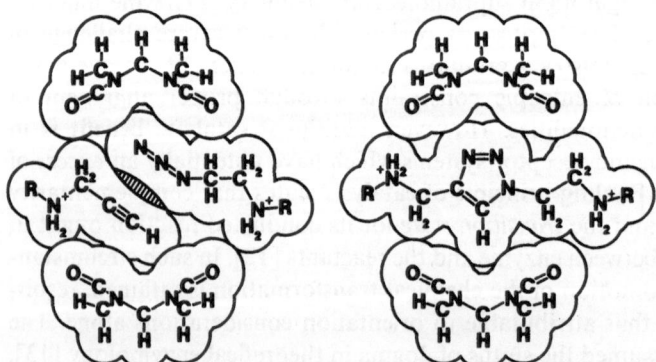

$$
\begin{array}{c}
HC{\equiv}CCH_2NH_3{}^+ \\
+ \\
N_3CH_2CH_2NH_3{}^+
\end{array}
\;\xrightarrow{\;k_0\;}\;
\underset{CH_2NH_3{}^+}{\overset{CH_2CH_2NH_3{}^+}{\text{triazole}}}
\;+\;
\text{triazole } C{-}CH_2NH_3{}^+
\tag{2}
$$

We found that a catalytic amount of cucurbituril markedly *accelerated* the reaction shown and rendered it regiospecific, yielding *only* the 1,3-disubstituted product. This result is explained by formation of a transient ternary complex between the reactants and the receptor. Simultaneous binding of both the alkyne and the azide, with one NH_3^+ coordinated to each set of carbonyls and with the substituents extending into the interior of cucurbituril, results in alignment of the reactive groups within the core of the receptor so as to facilitate production of the 1,3-disubstituted triazole. The proposed mechanism may be visualized with the aid of Fig. 7 (R = H).

Fig. 7. Conjectured cross-sectional representation of intermediate cycloaddition complexes: substrates + cucurbituril and product + cucurbituril (R = H or *t*-Bu). Outlines are drawn as for Fig. 4. *Shaded region* corresponds to strain-induced compression of substrates, promoting reaction (cf. Sect. 3.2); from [15] with permission

Rates for this reaction may easily be measured by disappearance of azide UV absorption. Most importantly, kinetic saturation behavior is noted; with sufficient amounts of the reactants cycloaddition velocity becomes independent of substrate concentration. As is familiar from enzyme catalysis, this indicates complete occupancy of all available cucurbituril by reacting species. In actuality, the rate of the catalyzed reaction under conditions of saturation was found to be the same as that for release of the product from cucurbituril. Such a stoichiometric triazole complex was independently prepared and its kinetics of dissociation were examined by the displacement technique previously outlined, giving the identical rate constant of $1.7 \times 10^{-4} \, \text{s}^{-1}$ under the standard conditions. (It is not uncommon for product release to be rate-limiting in enzymic reactions).

Since product dissociation is the slow step, no information about the actual cycloaddition stage is accessible from steady-state kinetics, However, by using a relatively large amount of cucurbituril, a "burst phase" at the initiation of reaction could be detected spectroscopically. This must correspond to "loading up" of the receptor by product, and consequently the rate of the induced cycloaddition itself could also be examined by careful data collection at early stages of reaction. Such measurements were complicated by "substrate inhibition", since evidently an adduct of cucurbituril with two propargylammonium ions forms competitively and this blocks out the azide, leading to a rate retardation in the presence of a large amount of the alkyne. However, with use of varying amounts of both reactants, one could by extrapolation obtain velocity estimates for the nominal bimolecular reaction between azidoethylammonium ion plus the noncovalent mono-adduct which has preformed between cucurbituril and propargylammonium ion. The second-order rate constant so obtained, value $6.3 \times 10^{-2} \, \text{M}^{-1} \text{s}^{-1}$, is suitable for comparison to k_0 in Eq. (2), for the uncatalyzed reaction. The resulting "catalytic acceleration" calculates to be a factor of 5.5×10^4.

The latter number incorporates just the chemical step(s) of formation of triazole within cucurbituril. Since the product release step apparently is at least 100-fold slower than the actual cycloaddition, the net *catalytic* acceleration should be adjusted downward by that amount. An instructive alternative estimation of kinetic enhancement is to compare the extrapolated limiting rate for cycloaddition within the complex (i.e. cucurbituril saturated with both reactants, $k_c = 1.9 \times 10^{-2} \, \text{s}^{-1}$) with the uncatalyzed unimolecular transformation of an appropriate bifunctional reference substrate as in Eq. (3) ($k_r = 2.0 \times 10^{-5} \, \text{s}^{-1}$). Such a comparison of first-order rate constants shows that the latter reaction is approximately a thousandfold slower than the cucurbituril-engendered transformation. This is attributable to necessity for freezing of internal rotational degrees of freedom that exist in the model system, which are taken care of when cucurbituril aligns the reactants, and concomitantly to an additional consideration which follows.

$$HC\equiv CCH_2NH_2{}^+CH_2CH_2N_3 \xrightarrow{k_r} \quad (3)$$

19

3.2 Argument for Bound-Substrate Destabilization

Current theory attempting to explain the uniqueness of enzymic reactions suggests that excess *potential binding energy* between substrate and enzyme may be diverted to the promotion of reaction [12, 13]. Ideally there should be a subtle misfit between the substrates and the active site, which is only relaxed as the catalyzed reaction proceeds to its transition state. In such circumstances the initially unrealized binding energy works to promote chemical transformation, by effectively lowering the net activation energy. We suspected that such a phenomenon might be occurring in the cucurbituril-induced cycloaddition.

In order to demonstrate the effect, it was felt that cleaner kinetic data was needed. This was secured by modification of the participants in the cycloaddition. A *tertiary* butyl group was introduced onto the amino nitrogen of each substrate (Fig. 7, R = *t*-Bu). This renders the reaction noncatalytic in the strict sense. Because the *tert*-butyl substituents are too large to pass through the cavity of cucurbituril, the resulting product of cycloaddition is a stable *rotaxane*, i.e. the triazole cannot dissociate. However, this is desirable in that the chemically meaningful presteady-state phase becomes delineated in a way that could only be incompletely realized in the previous kinetics. As a bonus the undesirable substrate inhibition by propargylammonium ion, which also obscured earlier kinetic measurements, could be avoided as well.

The resulting reaction profile fits the scheme shown in Eq. (4) (wherein C = cucurbituril, Z = azide, Y = alkyne, T = triazole). The individual (binary) dissociation constants for the reactants plus cucurbituril were independently

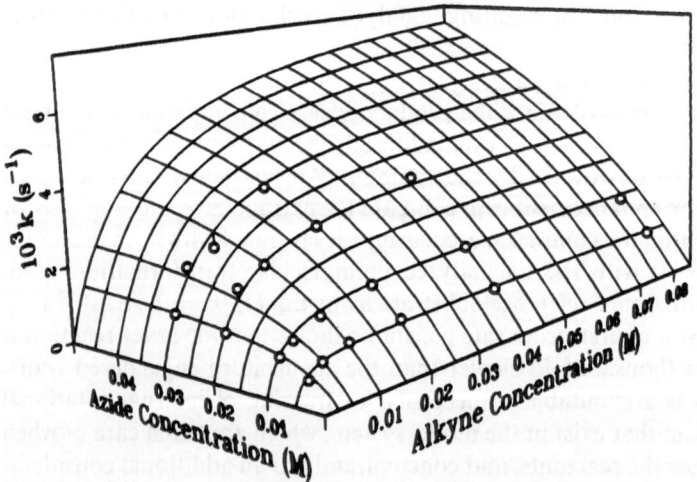

Fig. 8. Three-dimensional presentation showing simultaneous dependence of apparent rate of cycloaddition within cucurbituril upon substrate concentrations for both $HC\equiv CCH_2NH_2^+$-$C(CH_3)_3$ and $N_3(CH_2)_2NH_2^+C(CH_3)_3$. *Circles* are data points correlating with concentrations given by intersections of Michaelis curves in the grid; from [15] with permission

determined by previously described NMR competitive complexation experiments involving a reference standard: $K_1 = 0.44$ mM (for Y), $K_2 = 3.5$ mM (for Z). The substrate dependences for cycloaddition in the presence of cucurbituril are summarized in Fig. 8. Saturation kinetic behavior is noted for both substrates (most unambiguously for Y). The surface drawn by the grid lines corresponds to a simultaneous least squares fit of experimental rates to a mathematical relationship drawn from Eq. (4). The statistical refinement gives a value for k_c (the rate constant for unimolecular reaction at saturation with both substrates) of 0.023 s^{-1}, which is the same as for the previous case within experimental error. Also secured by curve fitting is a value for K_3, a Michaelis dissociation constant that corresponds to binding of azide (Z) to the pre-existing complex between cucurbituril and alkyne (C·Y), leading to the productive species (ternary complex, C·Y·Z). This parameter was found to have a value of 33 mM ($= K_3$). The ratio k_c/K_3 comprises a specificity constant (value 0.7 M^{-1}s^{-1}), which in this case indicates a sizable kinetic acceleration of 6×10^5-fold for the induced cycloaddition relative to the uncatalyzed reaction, as in the previous instance.

$$
\begin{array}{ccccc}
 & \overset{C \cdot Y}{\nearrow K_1} & & \overset{K_3}{\searrow} & \\
C & & & & C \cdot Y \cdot Z \xrightarrow{k_c} C \cdot T \\
 & \underset{K_2}{\searrow} Z & & \underset{C \cdot Z}{\nearrow} Y & K = K_1 \cdot K_2 / K_3
\end{array}
\tag{4}
$$

However, the more interesting number is the ratio between K_3 ($=[Z] \cdot [C \cdot Y]/[C \cdot Y \cdot Z]$) and K_2($=[Z] \cdot [C]/[C \cdot Z]$). It appears that K_3 is 75-fold larger than K_2. This means that it is correspondingly more difficult to incorporate an azide moiety into the cavity of cucurbituril when it is already occupied by alkyne, relative to when it is empty (i.e. a straightforward comparison of dissociation constants). Such negative cooperativity signifies that the presence of one reactant within the interior of cucurbituril diminishes the thermodynamic affinity of the second reactant for the receptor. This in turn implies that the cavity within cucurbituril is slightly undersized for optimal binding of both species at once. Consequently the substrates experience a *strain* when they are simultaneously incorporated, which tends to *compress* them together. The shading in Fig. 7 is intended to represent this. Common sense, as well as the confirmatory observation that 1,3-dipolar cycloadditions typically show a large negative volume of activation, indicate that the strain would tend to promote reaction. This should have additional kinetic benefits beyond that attributable to orientational effects alone, provided that the *transition state* for the cycloaddition were a better fit to the cavity within cucurbituril, as can readily be imagined.

The latter is of course the Pauling principle of catalysis. A factor of 75 could provide 2–3 kcal/mol reduction in the activation energy, assuming that most of the implicit strain is relieved in the transition state. This is also chemically plausible, since independent evidence (cf. Sect. 2.2.2) has suggested that the optimum size for a guest lies between that of a six-membered (benzene) and a five-membered ring (triazole). A deeper analysis suggests that the observed factor represents a lower limit [15]. The possibility cannot easily be excluded that K_3 corresponds primarily to *nonproductive* binding (with one of the reaction participants binding externally to cucurbituril, with only its ammonium ion coordinated to an occulus), and that the active complex shown in Fig. 7 is but a minor participant in a prefatory equilibrium involving productive and nonproductive complexes. In such an eventuality the strain activation could actually be much greater in the productive mode, but it would yield no additional kinetic benefit, since any increase in reaction velocity would be negated by a pre-equilibrium disfavoring the properly oriented substrates. That possibility has been explored by an inhibition study in which the putative nonproductive mode of coordination was simulated employing *neo*hexylamine. The results suggested that true productive binding was indeed being observed in the accelerated cycloadditions, but for details the original article should be consulted [15].

In summary, the proper criterion for measuring kinetic acceleration in catalyzed bimolecular reactions is a comparison of second-order rate constants, which entails knowledge of velocities for one substrate reacting with a saturated complex of the second substrate plus catalyst. In the present case a factor of approximately 10^5 was observed for the induced reaction itself. In enzymology these parameters (e.g. k_c/K_3) are known as *specificity constants*. It is relevant to note that any catalytic benefit arising from *reactant destabilization* as previously evinced for the cycloaddition is *not reflected* in the specificity constant, due to a cancellation of the opposing consequences for substrate binding (K_3) and for the velocity of the ensuing reaction (k_c). In order to produce a Pauling principle effect in this kinetic parameter, there must be unique, positive (attractive) intermolecular forces between transition state and catalyst, that are specifically absent in the initial complex of reactants. There is no autonomous method for experimental detection of such forces, in a manner analogous to that shown here for bound-substrate destabilization, although avenues are available to their investigation [16].

3.3 Homogeneous Polymerization

The sizeable catalysis noted in the previous sections suggested a model homogeneous polymerization induced by cucurbituril [17]. The *bis*propargylamine and *bis*azidoethylamine shown in Eq. (5) were prepared, in the hope that cucurbituril would link them into an alternating copolymer. An elegant sequential growth process can be envisaged, in which monomers are successively fed into one portal of cucurbituril, as polymer emerges from the other occulus. Alas, in practice the linkage operation shuts down after one or two cycloadditions. Evidently, the triazole moiety adheres too firmly to the interior of cucurbituril. In the simpler bimolecular reaction described earlier (cf. Sect. 3.1), the product dissociates cleanly from cucurbituril. The resulting gain in translational entropy is enough to ensure availability of catalyst for the next cycle. In the present instance cucurbituril must translocate intramolecularly from the previously synthesized ring to the connected alkyne or azide moiety, which constitutes a much weaker binding site (providing only one ion-dipole locus of attraction). The unfavorable partitioning between nonproductive and productive binding modes apparently greatly restricts the chain length which can be grown, with the available catalyst rapidly becoming tied up in an inert product complex. This is a problem to be born in mind in designing synthetic catalysts; some provision should be made for ensuring that product does not cumulatively inhibit the reaction.

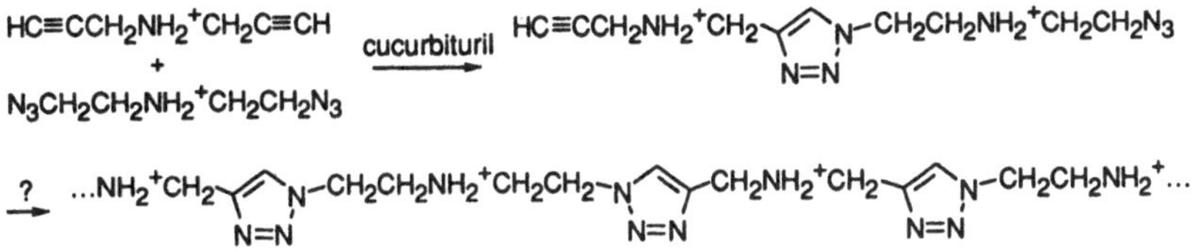

$$(5)$$

4 Conclusions

Studies with cucurbituril have been quite fruitful, both in providing an opportunity for systematic investigations of the thermodynamic and kinetic aspects of molecular recognition, and as a model for enzymic reactions. These studies have been immeasurably aided by its ready availability in reasonable quantity, in essentially two steps from urea, glyoxal, and formaldehyde. The extreme rigidity of the polycyclic cage structure of cucurbituril constitutes its most exceptional characteristic. This confers high selectivity among potential guests, allowing quantitative analysis of hydrophobic and other binding phenomena. This in turn has allowed rational applications, such as the molecular switch and substrates tailored for catalysis. We hope that our efforts may help to point the way for the emerging field of nanotechnology.

Acknowledgements. This work was supported at various stages by the University of Illinois Research Board, the Dow Chemical Co. Foundation, and the Office of Naval Research.

5 References

1. Behrend R, Meyer E, Rusche F (1905) Liebigs Ann Chem 339: 1
2. Freeman WA, Mock WL, Shih NY (1981) J Am Chem Soc. 103: 7367
3. Freeman WA (1984) Acta Crystallogr B40: 382
4. Shih NY (1981) Thesis, University of Illinois at Chicago (1982) Diss Abstr Int B42: 4701

5. Flinn A, Hough GC, Stoddart JF, Williams DJ (1992) Angew Chem Int Ed Engl 31: 1475
6. Mock WL, Shih NY (1983) J Org Chem 48: 3618
7. Mock WL, Shih NY (1986) J Org Chem 51: 4440
8. Mock WL, Pierpont J (1990) J Chem Soc Chem Commun 1509
9. Mock WL, Shih NY (1988) J Am Chem Soc 110: 4706
10. Hine J, Mookerjee PK (1975) J Org Chem 40: 292
11. Mock WL, Shih NY (1989) J Am Chem Soc 11: 2697
12. Pauling L (1948) Am Sci 36: 51
13. Jencks WP (1975) Adv Enzymol Relat Areas Mol Biol 43: 219
14. Mock WL, Irra TA, Wepsiec JP, Manimaran TL (1983) J Org Chem 48: 3619
15. Mock WL, Irra TA, Wepsiec JP, Adhya M (1989) J Org Chem 54: 5302
16. Mock WL, Freeman DJ, Aksamawati M (1993) Biochem J 289: 185
17. Adhya M (1986) Thesis, University of Illinois at Chicago (1986) Diss Abstr Int B47: 1056

Molecular Clips and Cages Derived from Glycoluril

Rintje P. Sijbesma[1] and Roeland J. M. Nolte[2]

[1] Laboratory of Organic Chemistry, Eindhoven University of Technology, P. O. Box 513, 5600 MB Eindhoven, The Netherlands
[2] Department of Organic Chemistry, NSR Center, University of Nijmegen, Toernooiveld, 6525 ED Nijmegen, The Netherlands

Table of Contents

Topics in Current Chemistry, Vol. 175
© Springer Verlag Berlin Heidelberg 1995

1 Introduction

The design and construction of hosts that are capable of selectively binding guest molecules requires precise control over geometrical features. This can be achieved by using versatile, rigid building blocks that allow the introduction of binding sites with directional binding interactions at well-defined positions. Among the building blocks frequently used (Fig. 1) are the cyclophanes [1], e.g. calixarenes [2], cyclotriveratrylenes [3], resorcinol-aldehyde tetramers [4], and others such as Kemps' triacid [5], Trögers' base [6], and the non-synthetic cyclodextrins [7].

Glycoluril **1** is a rigid, concave molecule, which can be easily functionalized via its four ureido nitrogen atoms. The two carbonyl groups in **1** are potential hydrogen bond acceptor sites. These features make **1** an excellent building block for synthetic host molecules. Glycoluril is the monomeric unit of cucurbituril,

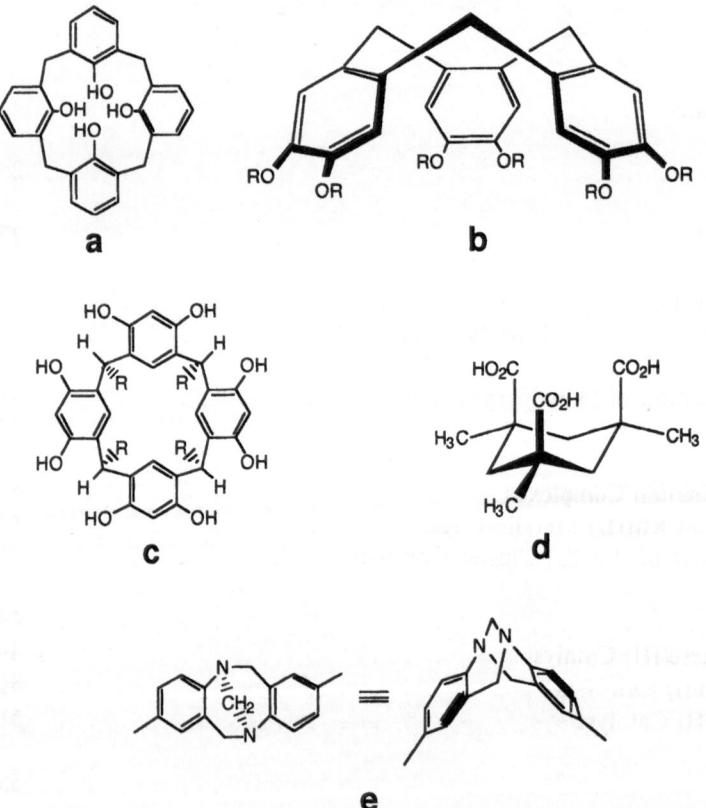

Fig. 1a–e. Examples of frequently used building blocks for synthetic host molecules: **a** calixarene; **b** cyclotriveratrylene; **c** resorcinol-aldehyde tetramer; **d** Kemps' triacid; **e** Trögers' base

a synthetic host which has been studied in depth by Mock and coworkers [8]. Over the last years, our research group has exploited the unique features of glycoluril in the development of a range of host compounds. In this article we will discuss these hosts and their use as molecular receptors, which are able to form well-defined complexes with species varying from alkali metal ions to neutral aromatic molecules. We will also describe efforts to develop synthetic catalysts that mimic the way enzymes accelerate chemical reactions.

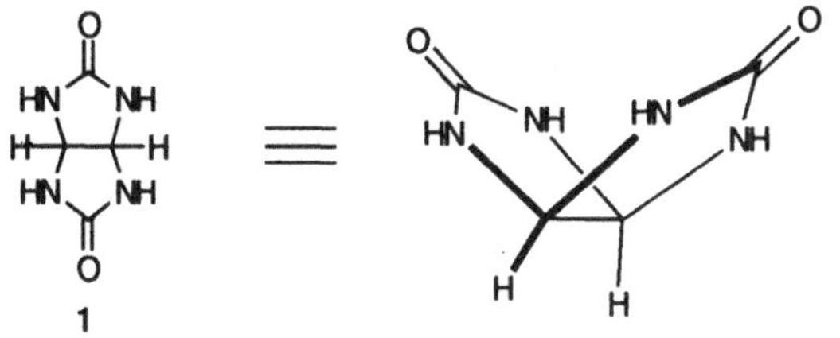

2 Molecular Clips

The two fused five-membered rings of glycoluril form a shallow cavity, which is further modified by adding two o-xylylene moieties. In addition, the convex side of the building block can be provided with two phenyl rings, allowing for optimal discrimination between the two faces of the glycoluril unit [9]. In this way, compounds of the general structure 2 are obtained. Their synthesis [10] is shown in Scheme 1. These compounds were expected to possess features that would make them good receptors for aromatic molecules, viz. two aromatic walls at a distance suitable for accommodating a benzene moiety in the cleft, and the possibility to have π-π stacking interactions with the guest.

The term "Molecular Clip" has been coined for molecules of type 2. That these molecules do indeed possess the geometric features of a clip is apparent from the X-ray structure of the tetramethoxy derivative 3a (Fig. 2) [11a, b]. The o-xylylene moieties of this molecule define a tapering cavity, the walls of which are at an angle of 39.5°, with the centers of the benzene rings 6.67 Å apart. The carbonyl groups of the glycoluril moiety, which are hydrogen-bond acceptor sites, are separated by 5.52 Å. It was also possible to obtain a crystal structure of the chiral dibromo-derivative 4 of clip 3 (Fig. 3). This compound was separated into its enantiomers by HPLC on a chiral stationary phase [12].

Complexation experiments show that compounds of type 2 are good receptors for dihydroxybenzenes, as expected [11]. Two types of interaction – hydrogen bonding and π-π stacking interactions – cooperate to bind a guest molecule in the cleft of the clip. Association constants for 3 and 5–7 with dihydroxy-substituted aromatic compounds were determined using ^{1}H-NMR titrations in $CHCl_3$ (Table 1). The role of π-π interactions is evident from the difference in K_a

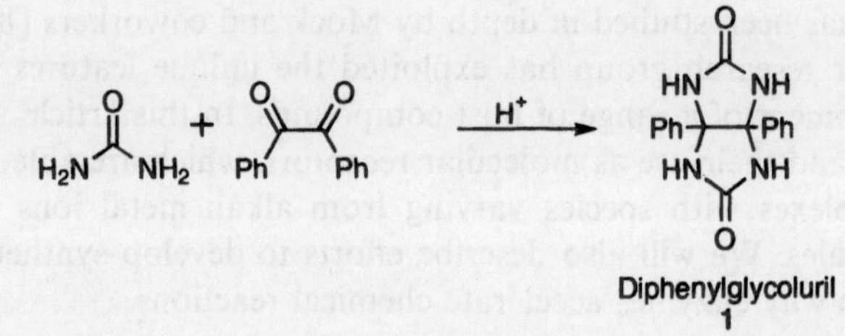

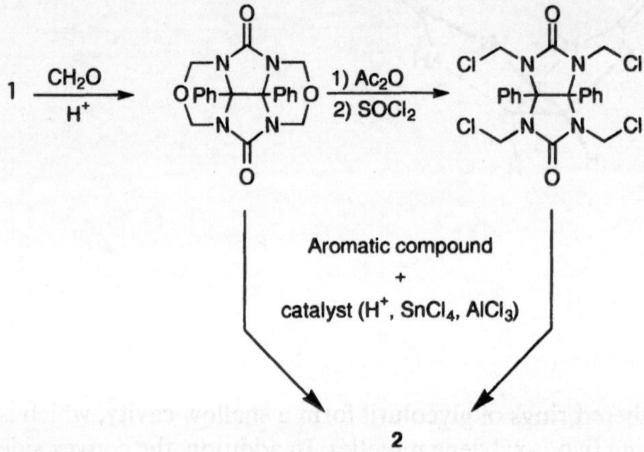

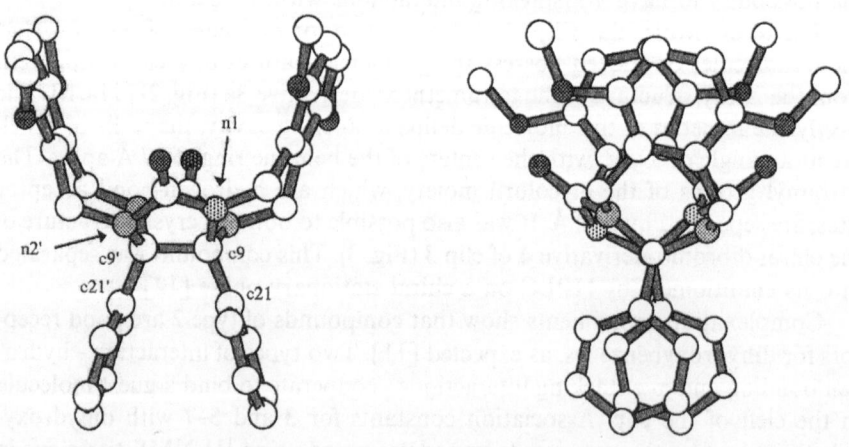

Diphenylglycoluril
1

Scheme 1

Fig. 2. X-ray structure of **3a**. (Reproduced with permission from the American Chemical Society)

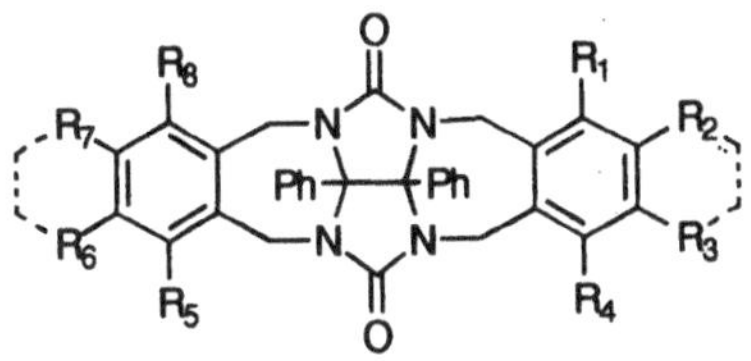

2

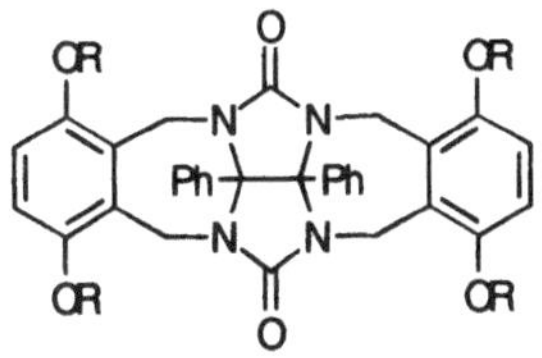

3

a R = Me
b R = H

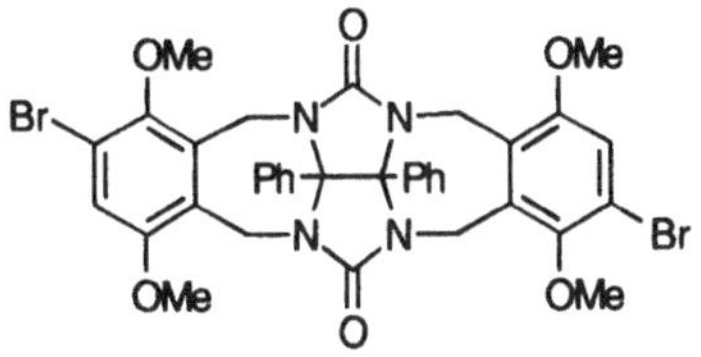

4

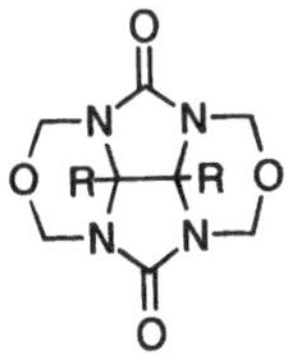

5

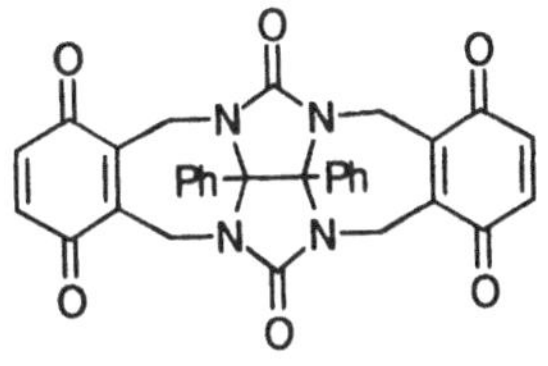

6

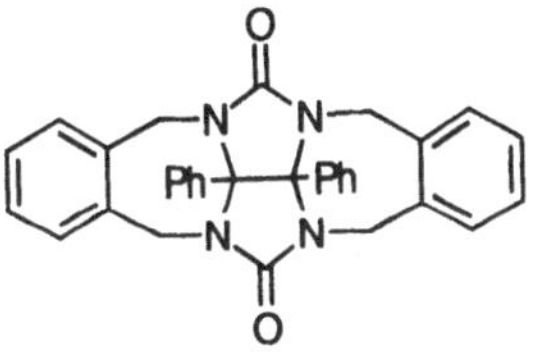

7

values of the complexes between model compound **5** with resorcinol and compound **3a** with resorcinol. The former compound can form two hydrogen bonds with resorcinol via its carbonyl oxygen atoms, but it lacks the cavity walls and, therefore, cannot further stabilize the complex by π-π interactions. Remarkably, the introduction of benzoquinone walls (see compound **6**), stabilizes the complex with resorcinol to a much smaller extent than the introduction of benzene or substituted benzene walls.

If the OH-groups of the guest become more acidic, e.g. upon changing the guest from resorcinol to 2,4-dibromoresorcinol, the K_a increases from 2600 to 5600 M^{-1} (Table 2, entries 2 and 3). In the case of 3,5-dihydroxybenzoic acid methyl ester, the K_a value reaches the relatively high value of 35000 M^{-1}

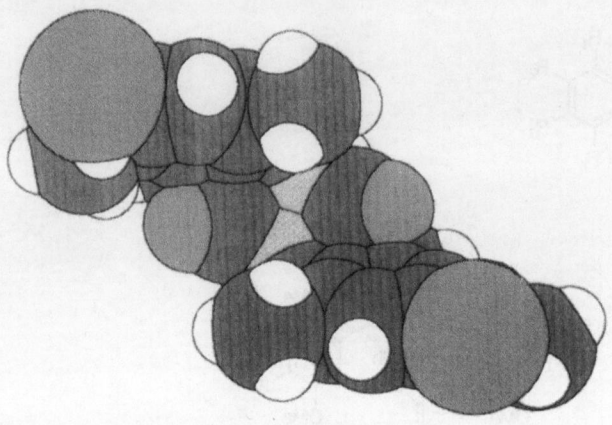

Fig. 3. X-ray structure of chiral clip **4**

Table 1. Association constants of host molecules with resorcinol in CDCl$_3$. T = 298 ± 2 K

Entry	Host	$K_a (M^{-1})$
1	**3a**	2600
2	**5**	25
3	**6**	30
4	**7**	200
5	**8**	<5

Table 2. Association contants of host **3a** with aromatic compounds in CDCl$_3$. T = 298 ± 2 K

Entry	Guest	$K_a (M^{-1})$
1	Catechol	60
2	Resorcinol	2600
3	2,4-Dibromoresorcinol	5600
4	2,7-Dihydroxynaphthalene	7100
5	3,5-Dihydroxybenzoic acid methyl ester	35000

(Table 2, entry 5). Due to the restrictions that the rigid host places on the orientation of the guest, the hydrogen bonds with dihydroxybenzenes are of a particular kind: they are formed with the π-electrons rather than with the n-electrons of the carbonyl groups (Fig. 4) [13]. This information can be obtained from the infrared spectra of the complexes. The OH-stretching vibrations corresponding to hydrogen bonds with the π-system are characterized by the fact that they are broader and less shifted from their original positions than the stretching vibrations of hydrogen bonds with the n-electrons [13a]. In the

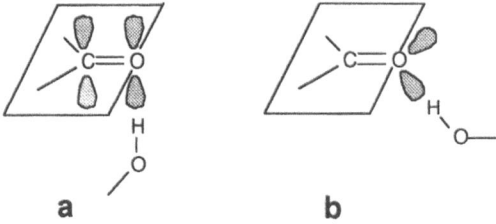

Fig. 4a, b. Two modes of hydrogen bonding with a carbonyl group: **a** with the π-electrons; **b** with the n-electrons. (Reproduced with permission from the American Chemical Society)

complex of phenol with **3a**, both kinds of vibrations are present, whereas in the 1:1 complex of the same host with resorcinol only the vibration for H-bonding to the π-system is visible.

The fixed distance between the carbonyl groups helps to confer binding selectivity to the hosts. The strength of the complexes with dihydroxy-substituted aromatic guests depends on the distance between the hydroxyl groups. The K_a's of the complexes of **3** with catechol, resorcinol and 2,7-dihydroxynaphthalene (Table 2, entries 1, 2 and 4) increase in the order given. In catechol, the distance between the OH-groups is somewhat too small to form a double hydrogen bonded complex. Moreover, catechol has an intramolecular hydrogen bond that must be disrupted to form such a complex. IR data suggest that catechol has only a single hydrogen bond with **3**, and consequently the K_a of the complex is low. In the complexes of **3** with resorcinol and 2,7-dihydroxynaphthalene two hydrogen bonds are present, but in the latter complex the distance between the OH-groups is such that more linear hydrogen bonds can be formed. This results in a higher K_a (7100 M^{-1}) of the complex with this guest than with resorcinol (2600 M^{-1}).

The proposed mode of binding of resorcinol in the clips was confirmed by comparing the experimentally derived ^{1}H-NMR shifts in the complex with shift values that were calculated using the Johnson and Bovey ring-current model [14]. Excellent agreement between these values was obtained for the structure depicted in Fig. 5. The main features of this structure are: (1) The hydrogen bonds between the carbonyl oxygen atoms of the host and the oxygen atoms in resorcinol have a length of approximately 2.7 Å; (2) the cavity walls of the host are somewhat closer together in the complex ($\pm$ 6.3 Å) than in the free host (6.7 Å, from the X-ray structure). This brings the cavity walls in Van der Waals contact with resorcinol. For 2,7-dihydroxynaphthalene, the shift calculations also indicated a symmetric mode of complexation with hydrogen bonds to each of the carbonyl oxygen atoms of the host. In the complexes with catechol, the induced shifts were rather small. Consequently the calculations could not be used to discriminate between the different modes of complexation, although good agreement between calculated and experimental shifts was obtained for a symmetric complex with catechol bound inside the cleft.

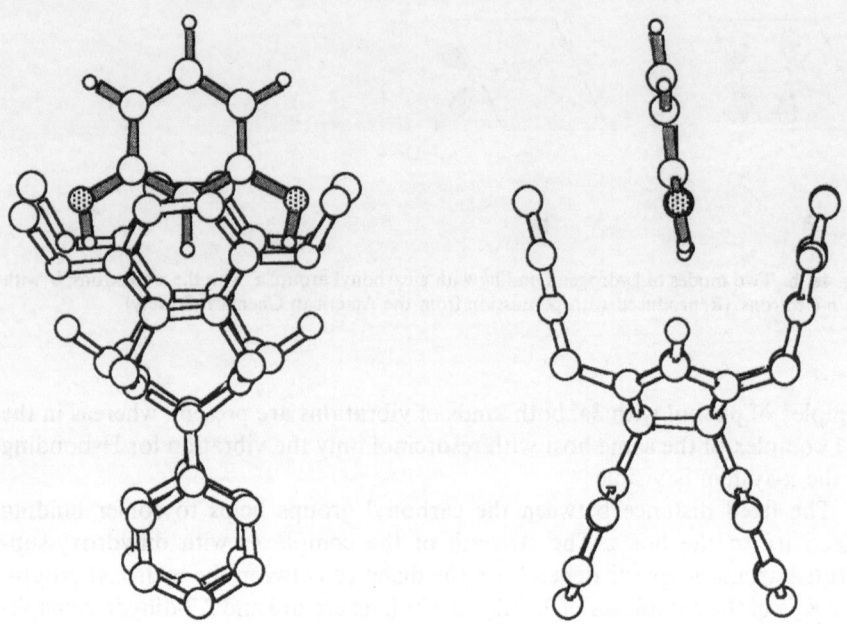

Fig. 5. Side and front view of the complex between resorcinol and **3a** based on NMR and X-ray data

In order to increase the π-π interaction with the aromatic guests, two molecular clips with naphthalene walls (compounds **8** and **9**) were synthesized. These molecules had quite unexpected properties. Surprisingly, the clip with 1,4-dimethoxynaphthalene cavity walls (**8**) did not bind guest molecules [15]. The reason for this behavior became clear when the X-ray structure was solved [16] (Fig. 6). All four methoxy groups of **8** were found to point into the cavity, completely blocking the access of a potential guest to the carbonyl groups. A similar structure may also be present in solution.

The receptor with 2,7-dimethoxynaphthalene rings (**9a**), exists in solution as a mixture of three conformers [17] (Fig. 7). Due to the fact that each of the cavity walls can have two orientations, an *anti-anti* (*aa*), a *syn-syn* (*ss*), and a less symmetric *syn-anti* (*sa*) conformation are possible. These three conformations

8

9a R = OMe
9b R = OAc

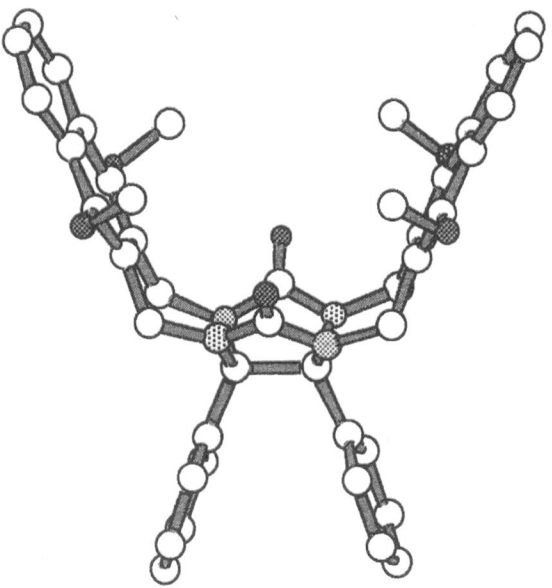

Fig. 6. X-ray structure of **8**. (Reproduced with permission from the Royal Chemical Society, London)

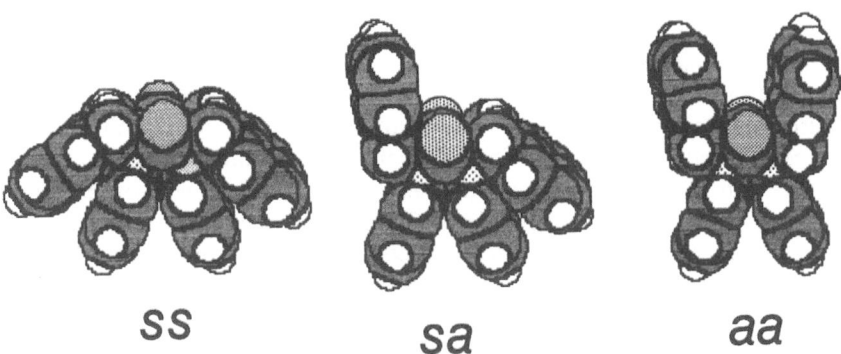

ss   *sa*   *aa*

Fig. 7. The three conformers of **9a**. (Reproduced with permission from the American Chemical Society)

interconvert slowly on the NMR time scale. The rate constants for the different interconversion processes in the tetraacetoxy derivative **9b** were determined by two-dimensional ^{1}H-NMR. It was concluded that the conformers interconvert by a single naphthyl flip process, and that processes in which two naphthyl groups flip simultaneously do not occur (Fig. 8).

Of the three isomers, only the *aa* conformer is able to stabilize complexes with aromatic guests through π-π stacking interactions with both naphthyl

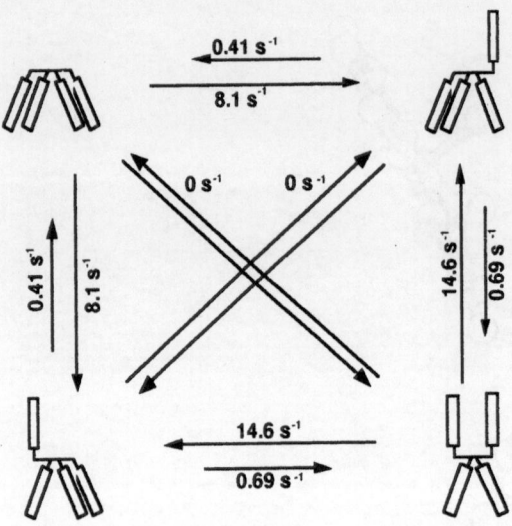

Fig. 8. Interconversion processes between the conformers of **9b**. (Reproduced with permission from the American Chemical Society)

groups. Consequently, the addition of π-electron poor aromatic substrates to a chloroform solution of **9** causes the relative amount of the *aa* conformer to increase. This phenomenon is reminiscent of Koshland's induced fit model for binding of substrates in enzymes [18]. The shift in conformational equilibrium was used to calculate the association constants of different guests with **9a**. Some of these association constants were also obtained by monitoring the charge transfer bands of the host-guest complex as a function of the guest concentration in UV titration experiments (Table 3). These experiments show that the *aa* conformation of the naphthalene clip selects the guest not only on the basis of the guest acceptor strength, but also on its size. In particular when the glycoluril framework of the clip interferes with substituents on the guest, the latter is prevented from taking a position optimal for π-π stacking with the cavity walls. This is very clearly seen in the case of trinitrotoluene, which does not bind in the cavity of **9a**, whereas the complexes with the weaker acceptors 1,2-dinitrobenzene and 1,3-dinitrobenzene have appreciable K_a values viz. 95 and 115 M^{-1} (Table 3, entries 3–6). Steric effects also play a role in the complex of **9a** with the very strong acceptors tetracyanoquinodimethane and tetracyanoethylene. These compounds form charge transfer complexes with the clip, but are not bound in its cavity, as concluded from the fact that their presence does not influence the conformational equilibrium.

Apart from being a receptor for aromatic molecules, **9a** also binds silver (I), as was demonstrated by a change in the conformational equilibrium upon addition of silver perchlorate to a solution of this clip.

Table 3. Apparent K_a values of the *aa* conformer of **9a** with aromatic guests in $CDCl_3$ at 298 K[a]

Entry	Guest	$K_a(M^{-1})$
1	Toluene	0
2	Nitrobenzene	5.5[b]
3	1,2-Dinitrobenzene	95
4	1,3-Dinitrobenzene	115[c]
5	2,4,6-Trinitrotoluene	0
6	1,4-Dicyanobenzene	185
7	Tetracyanoquinodimethane	0
8	Tetracyanoethylene	0

[a] Estimated error in K_a 15%. [b] In $CDCl_3$/nitrobenzene 1 : 1 (v/v).
[c] Determined from an UV-titration in $CHCl_3$, estimated error in K_a 10%.

The above mentioned experiments show that guests are preferentially bound by the *aa* conformation of **9a**. In Sect. 3 we will describe how this concept can be used to construct a molecular system that switches between strongly and weakly binding states.

3 Molecular Baskets

The complexation of guests in the molecular clips is achieved through hydrogen bonding and π-π stacking interactions. In order to extend the number and diversity of the binding interactions and consequently the range of substrates that can be bound, the clips were functionalized with crown ether fragments. The resulting compounds (**10–19**) have the shape of a basket (see Fig. 9).

The basket compounds were synthesized by reacting compound **3b** with oligoethyleneglycol dichlorides in DMSO [19] (e.g. see Scheme 2). These reactions have remarkably high yields (up to 75%) for closure of 22- to 25-membered rings. Presumably a template effect of the cation of the base used in the reaction is operative. Basket compounds with a nitrogen atom in the crown ether moieties are most conveniently synthesized by the route depicted in Scheme 3 [20]. With this method even higher yields (up to 89%) were obtained in the ring closure steps. Using similar routes, the chiral baskets **20–22** were also prepared. Compounds **20** and **21** were resolved into enantiomers by chromatographic procedures.

3.1 Complexation of Alkali Metal Ions

The basket compounds contain two crown ether moieties, which are good binders of alkali metal ions. It was therefore expected that these compounds

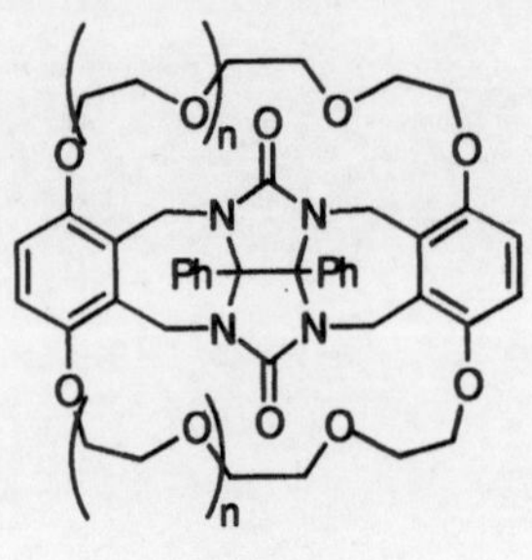

10 n = 1
11 n = 2
12 n = 3

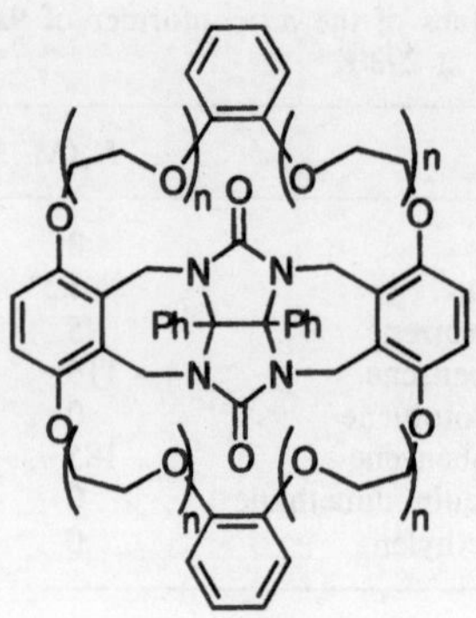

13 n = 1
14 n = 2

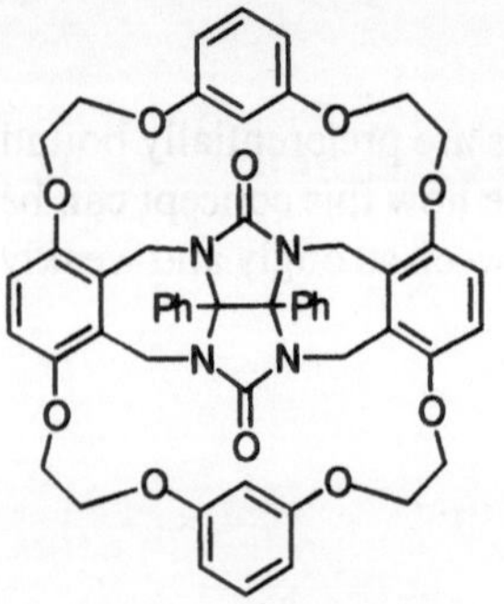

15

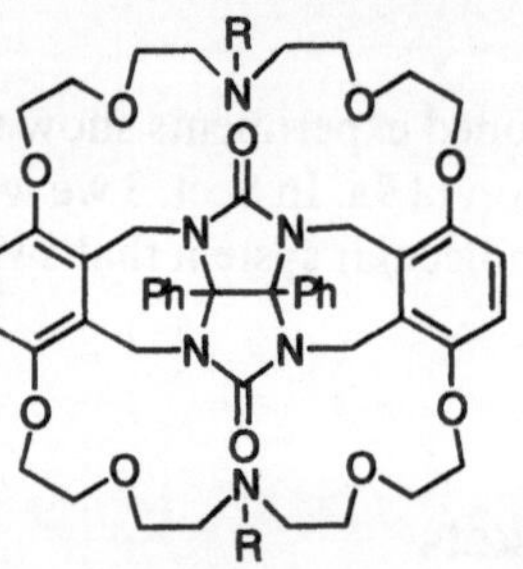

16 R = benzyl
17 R = H

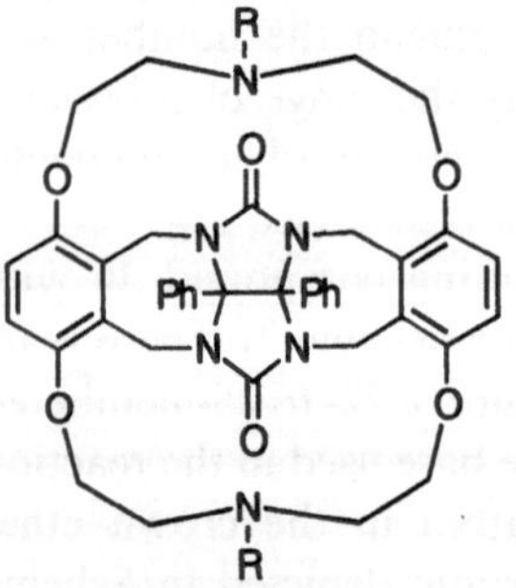

18 R = benzyl
19 R = H

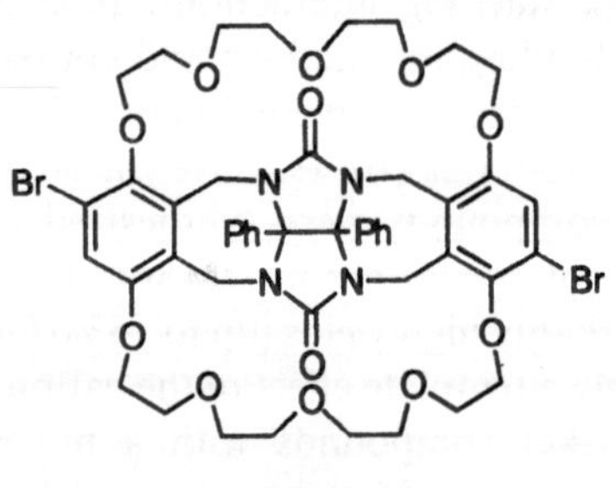

20

would bind metal ions in a 1:2 host-guest stoichiometry. The association constants of complexes between hosts **10–12** and a number of alkali metal ions were determined by the picrate extraction technique [19]. Binding stoichiometries were calculated from ^{1}H-NMR titrations in $CDCl_3/DMSO$-d_6 with potassium picrate or potassium thiocyanate as guest molecules. Surprisingly, a 1:2 host-potassium complex stoichiometry was only found in the case of the host with the longest oxyethylene bridging groups (**12**). Hosts **10** and **11** only bind

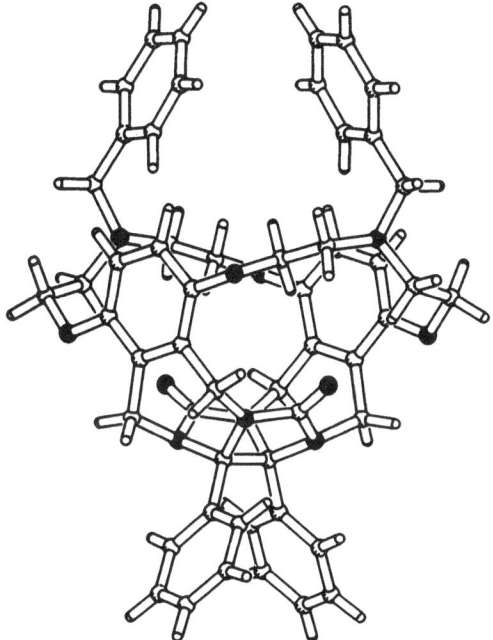

21 $[\alpha]_D^{20} = (+)$ and $(-)$ 16.4°

(S,S)-**22** $[\alpha]_D^{20} = +15.6°$

Fig. 9. X-ray structure of a molecular basket containing two benzylaza crown ether fragments (compound **18**). (Reproduced with permission from the Royal Netherlands Chemical Society)

one potassium ion. To explain the unexpected stoichiometries the formation of a clamshell complex was proposed (Fig. 10). The shorter crown ether bridges in **10** and **11** do not have enough oxygen atoms available to fully solvate two potassium ions. Moreover, these bridges are too short to allow the carbonyl oxygen atoms of the glycoluril unit to be involved in the binding process. As

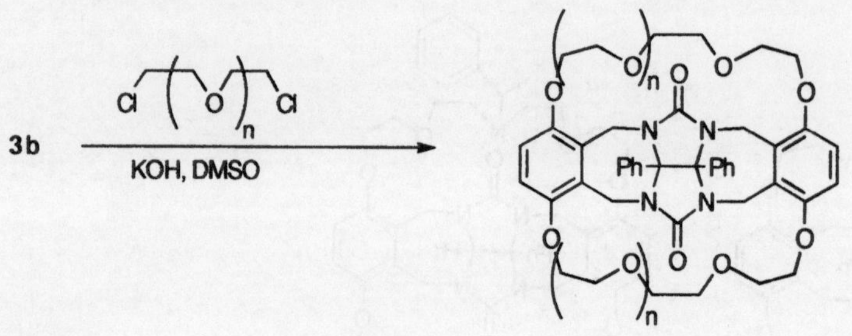

3b ─── KOH, DMSO ───⟶

Scheme 2

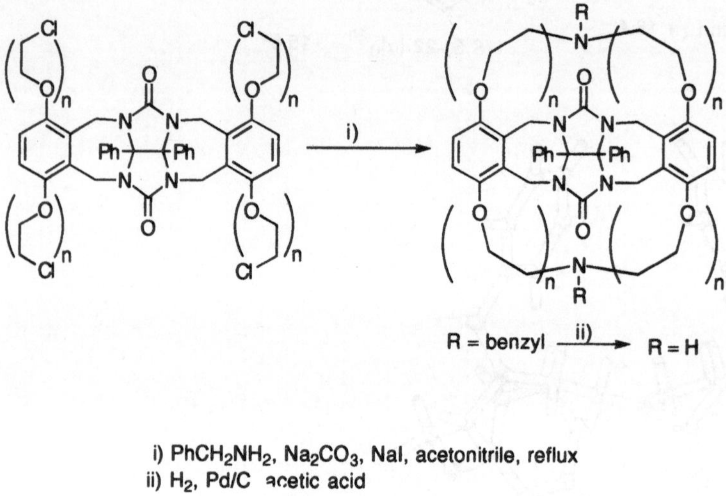

R = benzyl ──ii)─⟶ R = H

i) PhCH₂NH₂, Na₂CO₃, NaI, acetonitrile, reflux
ii) H₂, Pd/C acetic acid

Scheme 3

a result the two crown ethers prefer to bend toward each other and to encapsulate only one metal ion. This explanation was confirmed when titration experiments with basket **20** were performed [21]. In this molecule the bridges cannot fold over the cavity, due to steric interference with the two bromo substituents on the cavity walls. Contrary to **10** and **11**, this host does form complexes with potassium in a 1:2 stoichiometry, although of lower stability.

The binding profiles of hosts **10–12** show the same pattern as those of the corresponding benzo-crown ethers (Table 4). In most cases the baskets are better binders than the crown ethers. Compounds **13** and **14** have o-phenylene moieties in their crown ether bridges. They display a binding behavior similar to that of **10–12**, except for the ammonium ions, which are bound stronger in the baskets without these groups [22]. In compound **15** two ethyleneoxy units are replaced by m-phenyleneoxy groups. This decreases the free energy of binding by 1.3 to 8.0 kJ/mol as compared to **10**.

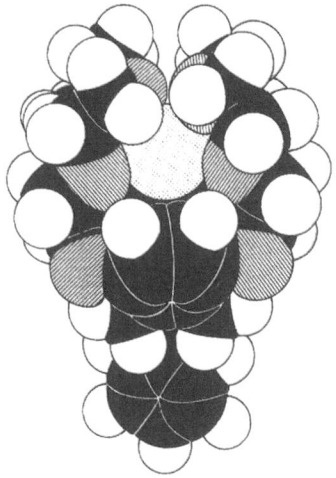

Fig. 10. CPK model of the proposed structure of the complex of **11** with an alkali metal ion ("clamshell complex"). (Reproduced with permission from the American Chemical Society)

Table 4. Free energies of binding of picrate salt guests to hosts at 298 K in CHCl₃ saturated with H₂O

Cation of guest	Host 10	Benzo-15-crown-5	11	Benzo-18-crown-6	12	Benzo-21-crown-7	13	14	15	16	17
						$-\Delta G°\,(\mathrm{kJ\cdot mol^{-1}})$					
Li⁺	30.6	30.6	28.1	26.8	27.7	25.1	31.8	27.2	23.9	31.8	35.2
Na⁺	33.1	35.2	37.7	33.1	32.7	32.3	33.1	31.0	26.4	33.9	35.6
K⁺	29.7	33.1	48.2	41.9	39.8	35.6	31.4	36.5	26.4	36.0	35.6
Rb⁺	27.7	29.3	42.3	38.5	37.3	37.3	28.9	35.6	27.7	34.4	35.2
Cs⁺	27.7	26.0	39.0	35.2	37.3	37.7	29.3	35.2	27.7	35.1	36.0
NH₄⁺	24.3	25.6	39.8	37.7	36.9	35.2	29.7	33.1	26.4	37.7	43.2
MeNH₃⁺	26.4		33.9		36.0		28.9	26.4	23.9	31.0	34.4
t-BuNH₃⁺	≪ 21		23.0		23.0		< 21	< 21	< 21	23.9	28.9

3.2 Complexation of Diammonium Salts

The presence of two crown ether moieties in the baskets allows the simultaneous binding of two ammonium groups. Organic diammonium salts of type $^+H_3N(CH_2)_nNH_3^+$ are bound very strongly, as is evident from the high association constants in Table 5 [19].

The complex of **11** with the short chain guest propanediammonium picrate has a relatively low K_a because the guest cannot stretch in the cavity and coordinate to both crown ether rings simultaneously. For the longer chain picrate salt ($n = 4$), the crown ether bridges must fold slightly toward each other

Table 5. Free energies of binding of diammonium dipicrate salts to hosts at 298 K in CHCl$_3$ saturated with H$_2$O

Cation of guest ^+H_3N-$(CH_2)_n$-NH_3^+	$-\Delta G^\circ$ (kJ$\cdot$mol^{-1}) Host 11	12	17
$n = 3$	39.8	37.7	51.1
$n = 4$	47.3	41.1	52.8
$n = 5$	52.8	42.7	49.9
$n = 6$	55.7	48.2	50.3
$n = 7$	55.7	52.4	50.3
$n = 8$	55.7	57.4	50.3
$n = 9$	56.6	57.4	50.3
p-xylylenediammonium	53.6	46.9	59.5
m-xylylenediammonium	53.2	46.1	57.4
o-phenylenediammonium	47.8	44.4	56.1
o-phenylenediammonium	51.5	50.3	71.6

to attain complexation of both ammonium groups. In the case of $n = 5$–6, the association constant reaches a maximum as the guest fits in the cavity of the host without much conformational adjustment of either partner. If the $(CH_2)_n$ chain is made longer, the K_a drops just slightly, because the guest molecule can accomodate the excess length by bending its $(CH_2)_n$ chain. A schematic representation of the various binding modes is shown in Fig. 11. The same binding trends are observed for **12**, except that the maximum association constant has in this case shifted to a longer chain guest ($n = 8$–9), due to the longer distance between the crown ether bridges in **12** as compared to **11**. The general mode of complexation as proposed in Fig. 11 was confirmed by ^{1}H-NMR studies. The protons of the CH$_2$ groups of the guests are shielded to varying degrees by the aromatic cavity walls of the host. For example, the induced shifts on the central methylene protons are low in the complex of **11** with tetramethylenediammonium picrate ($n = 4$), they reach a maximum for the complex with $n = 5$, and decrease again if $n > 6$.

A solution containing both **11** and hexamethylenediammonium picrate shows separate signals for the protons of free and bound guest at room temperature, indicating a slow exchange process between these two species on the NMR time scale. The binding kinetics of complexes of **11** and heptamethylene- and pentamethylene diammonium picrates are such that broad peaks are obtained at room temperature, whereas for the complexes of **11** with $^+H_3N(CH_2)_nNH_3^+$ ($n = 3$, 4, 8 or 9), the peaks are sharp, suggesting that the exchange of free and bound guest is fast on the NMR time scale.

Compounds **11** and **12** also form complexes with aromatic diammonium salts (Table 5). The ΔG values of binding vary from 44–54 kJ/mol. In the case of p-xylylenediammonium picrate, the aromatic protons of the guest give two signals, because two of them are in the shielding zone of the cavity walls, and the other two are in the deshielding zone.

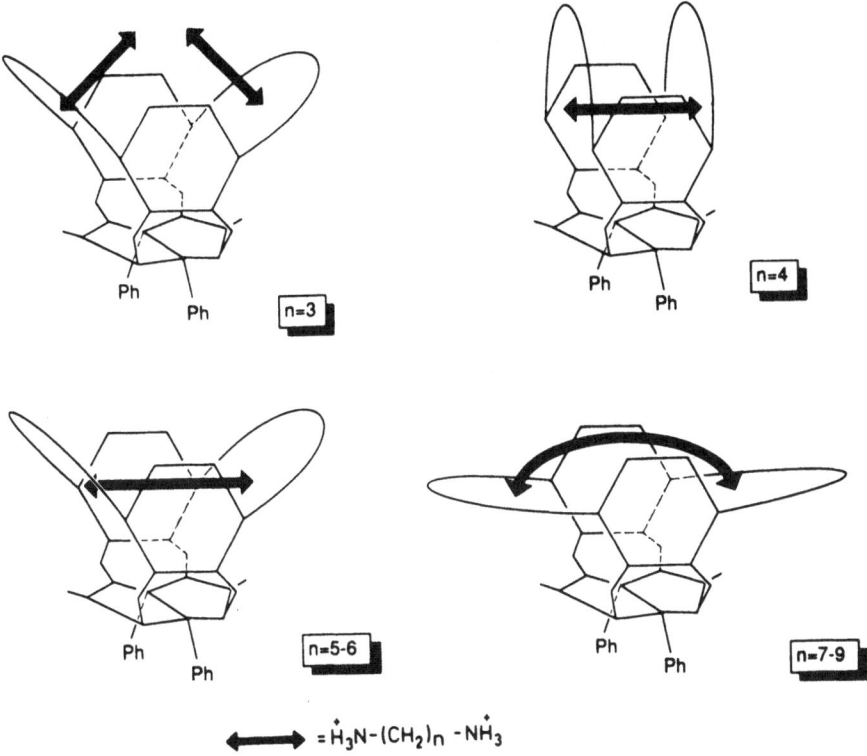

Fig. 11. Schematic representation of complex formation between picrate salts of $^+H_3N(CH_2)_nNH_3^+$ (n = 3–9) and **11**. (Reproduced with permission from the American Chemical Society)

The association constant of **17** with tetramethylenediammonium picrate is significantly higher than that of **11** with this guest [23]. This phenomenon is caused by proton transfer from the guest to the host, as was deduced from NMR chemical shift changes that occur upon complexation. Electrostatic repulsion between the two ammonium groups is released when a proton is transferred to a nitrogen atom of **17**. The same effect is responsible for the extremely high association constant of o-phenylenediammonium picrate with **17**. This complex has a free energy of binding as high as 71.6 kJ/mol.

3.3 Complexation of Dihydroxybenzenes

The nitrogen atoms in the bridges of basket compounds **16–19** are good hydrogen bond acceptors. They posseses a higher basicity than the carbonyl oxygen atoms of the glycoluril units, which were used as acceptor sites in the complexes of the molecular clips with dihydroxybenzenes. It was expected that the possibility to tune the position and orientation of the nitrogen atoms by adjusting the length of the bridges would endow each of the basket hosts to selectively bind a particular dihydroxybenzene derivative. The complexation of various dihydroxybenzenes in the cavities of **16–19** was investigated by ^{1}H-NMR titration experiments [20]. The K_a values of the complexes of catechol and resorcinol with **16** are similar to those of the complexes with **3a** (Table 6). Hydroquinone, which is not bound in the cavity of **3a**, does form a complex with **16** ($K_a = 650$ M^{-1}). The former host is unable to form two hydrogen bonds with hydroquinone, whereas **16** can use its nitrogen atoms to achieve binding. Modifying the guest with electron withdrawing groups increases the association constant, because the hydroxyl groups can function as better hydrogen bond donors. The association constant of **16** with 2,3-dicyanohydroquinone is one of the highest reported in the literature: $K_a = 3 \times 10^5$ M^{-1}.

The value of the association constant of the complexes between **16** and bromo and chloro substituted hydroquinones depends on three factors: hydrogen bond strength, steric hindrance and the occurrence of intramolecular hydrogen bonding within the guest. The K_a's measured for 2-bromo- and 2-chloro-hydroquinone are higher than that measured for hydroquinone. In the complexes of the former guests the hydrogen bonds are stronger than in the complex of the latter guest. Upon binding, one intramolecular hydrogen bond in the guest must be broken, which reduces the free energy of complexation. Nevertheless, the net effect is an increase of K_a, viz. from 650 to 1500 M^{-1} (2-chlorohydroquinone) and to 1800 M^{-1} (2-bromohydroquinone). In 2,3-disubstituted halogenohydroquinones, both OH groups are intramolecularly hydrogen bonded. The accompanying loss in complexation energy results in association constants that fall in between the values of the monosubstituted derivatives and the value of hydroquinone. The 2,5-disubstituted derivatives cannot enter the cavity of the host because of steric hindrance: the guest has a bulky substituent on each side of the hydroquinone ring, which precludes complexation.

Table 6. Association constants and free energies of complexation for complexes of **16** with dihydroxybenzene derivatives in CDCl$_3$ at 298 $\pm$ 2K

Entry	Dihydroxybenzene	$10^{-2} \times K_a{}^a$	$-\Delta G°$ (kJ·mol^{-1})
1	Catechol	0.70	10.5 ± 0.1
2	4,5-Dibromocatechol	12^b	17.6 ± 0.2
3	Resorcinol	29	19.7 ± 0.1
4	Hydroquinone	6.5	16.0 ± 0.2
5	2-Chlorohydroquinone	15^b	18.0 ± 0.2
6	2-Bromohydroquinone	18	18.5 ± 0.1
7	2,3-Dichlorohydroquinone	7.2^b	16.3 ± 0.2
8	2,5-Dichlorohydroquinone	$-^c$	–
9	2,3-Dibromohydroquinone	14	17.9 ± 0.1
10	2,5-Dibromohydroquinone	$-^c$	–
11	2,3-Dicyanohydroquinone	3000^d	31.2 ± 1.0

^{a} Estimated error in K_a 4%, unless otherwise indicated. ^{b} Estimated error 10%. ^{c} Induced shifts are too low to determine K_a. ^{d} This value was determined by a solid-liquid (CDCl$_3$) extraction experiment, estimated error in K_a 50%.

Baskets **18** and **19** were especially designed to bind catechol and catechol derivatives. The distance between the crown ether nitrogens in these host compounds is such that each of these atoms can form a hydrogen bond with a guest hydroxyl group. It was expected that the reduced flexibility of the bridges in **18** and **19** as compared to **16** would further promote guest binding. The results obtained with **18** were quite disappointing. Catechol, as well as resorcinol and hydroquinone were bound with association constants not exceeding 50 M^{-1}. The debenzylated analog **19**, however, behaved as predicted by displaying a $K_a = 500$ M^{-1} for catechol, which is 12.5 times larger than the K_a of **17** with this guest. In order to understand the reasons for the low binding affinities of **18**, an X-ray structure of this host was determined [24] (see Fig. 9). This structure reveals the features known to be required for complexation of dihydroxy-substituted aromatic guests: a cavity flanked by benzene rings with a geometry similar to that found in the X-ray structure of compound **3a**, and two crown ether nitrogen atoms with their non-bonded electron pairs pointing into the cavity. The reason that **18** is not a good host comes from steric hindrance: the benzyl groups attached to the nitrogen atoms point upward and shield the entrance of the cavity. ^{1}H-NMR NOE experiments show that in solution the above-mentioned conformation is at least partly preserved.

3.4 Cooperative Binding

In Sect. 2 we described a molecular clip (**9a**) that changes conformation upon binding a guest molecule. Compound **23** is a bis-aza-crown ether derivative of this clip. Similar to **9a**, it exists mainly in the *sa* conformation which is unfavourable for binding aromatic molecules [25]. Interestingly, when a sodium or potassium ion is added to a solution of **23**, the molecule changes its conformation to the *aa* form. The driving force is the fact that the crown ether rings interact more strongly with alkali metal ions when both walls of **23** are pointing upward. The possibility to change conformation causes **23** to display *cooperative binding* of alkali metal ions. Binding of the second ion is stronger than that of the first ion because the crown ether rings have already been brought into the correct conformation for complexation. The magnitude of the

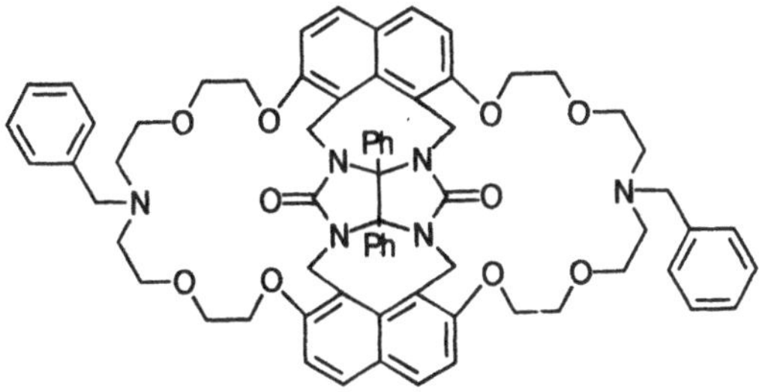

23

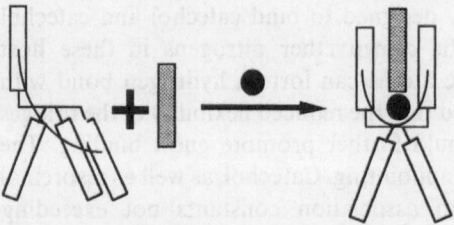

Fig. 12. Induced binding of a guest in **23** by the addition of a metal ion. (Reproduced with permission from the American Chemical Society)

cooperative effect was determined by ^{1}H-NMR titration experiments in which the fraction of molecules of **23** that are in the ion-binding *aa* conformation is monitored as a function of the number of equivalents of salt that is added. Fitting the theoretical binding curves to the titration data showed that a second K^+ ion is bound approximately 100 times more strongly than the first ion. The increased affinity for a second guest in **23** is called an *allosteric effect*, which is an important regulatory mechanism in biological systems.

The ^{1}H-NMR spectra of the complexes of **23** with alkali metal ions show remarkable differences. In the complex with sodium ions, the benzyl groups are in the shielding zone of the naphthalene walls, which suggests that these groups are bound in the cavity of the molecule. The potassium complex has a more open structure in which the cavity is not occupied by benzyl groups.

Addition of an alkali metal ion to **23** should also make this host molecule a better binder of aromatic guest molecules. In the *aa* conformation unlike the *as* conformation-the π-electron rich naphthalene walls are capable of sandwiching π-electron poor molecules (see Fig. 12). Indeed, upon addition of a potassium salt to a solution of **23**, the K_a for binding of 1,3-dinitrobenzene increases by a factor of 2 to 6, depending on the solvent systems [26]. Consistent with the proposed structure of the complex of **23** with sodium ions, addition of these ions has no effect on the K_a of **23** with 1,3-dinitrobenzene.

4 Cage Coordination Complexes

4.1 Pd(II) and Rh(III) Metallo-cages

The concave framework of diphenylglycoluril has been used to construct metallo-cages [27, 28]. To this end the molecule has been provided with four alkyl or oxyethylene chains terminating in ligating groups such as (benz)imidazolyl or pyridyl groups. On coordination to a transition metal ion this tetrapodal ligand system forms a cage, which is schematically depicted in Fig. 13. The phenyl groups on the convex side of the glycoluril unit increase the

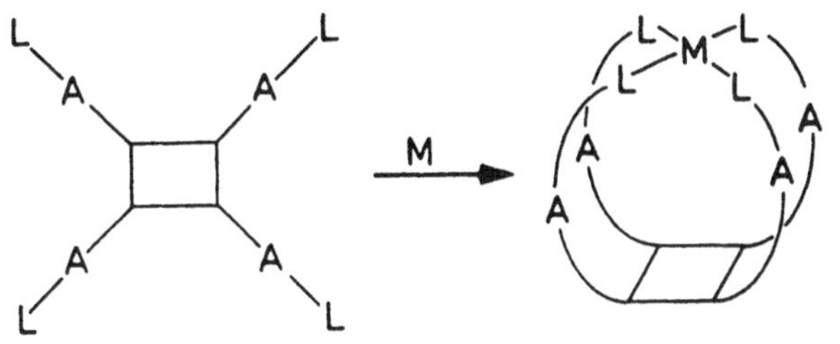

Fig. 13. Metallo cage strategy

solubility of the complex and direct the coordination process toward the concave side of the molecule.

The palladium(II) complex of **24** was prepared by adding 1 equivalent of [Pd(CH$_3$CN)$_2$Cl$_2$] to a solution of the ligand in methanol [27]. The complex was shown by conductivity measurements to be a 1:2 electrolyte, in accordance with the structure [Pd(**24**)]Cl$_2$. Spectroscopic data indicate that all four imidazolyl groups are coordinated to the Pd(II) ion, suggesting a cage-like

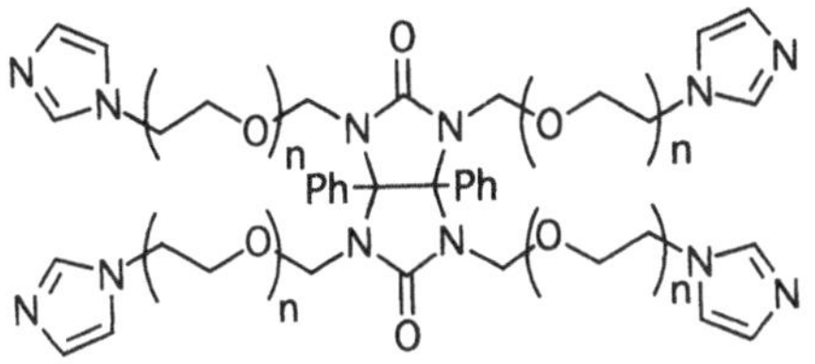

24 n = 1
25 n = 2
26 n = 3

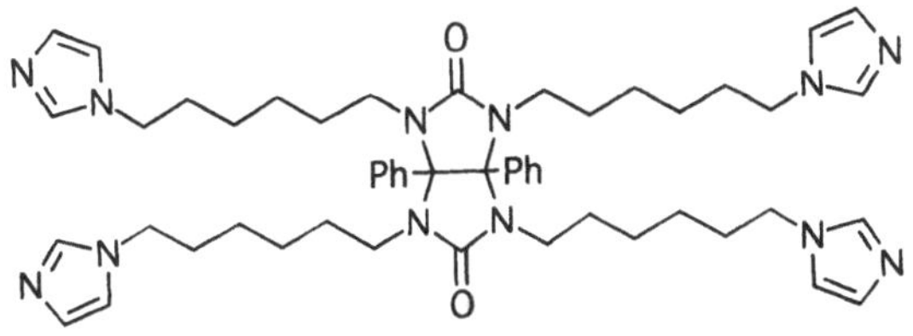

27

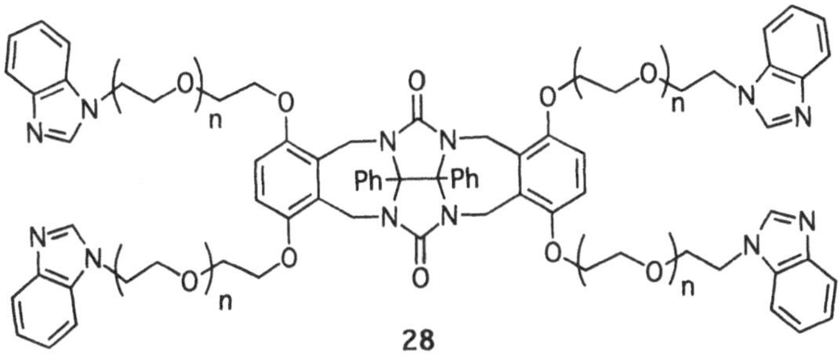

28

structure for this compound. However, ^{1}H-NMR experiments carried out at various temperatures and NOE measurements show that this picture is too simple: the cavity of [Pd(**24**)]Cl$_2$ is unstable and collapses via a twisting motion of the arms and the imidazolyl groups (Fig. 14). At room temperature a fast equilibrium exists between a left-handed and a right-handed twisted complex. At lower temperatures the dynamic equilibrium slows down, and at $-95\,°C$ two separate imidazolyl H^2 proton signals can be observed in the ^{1}H-NMR spectrum due to non-equivalence of the imidazolyl groups in a helically twisted structure.

Ligands **25–27**, which have their imidazolyl groups connected to the diphenylglycoluril units via $CH_2O(CH_2CH_2)_2$, $CH_2O(CH_2CH_2)_3$, or C_6H_{12} spacers, form complexes with palladium(II) of the type shown in Fig. 15. Spectroscopic data indicate that these compounds also display fluxional behavior. At any one time only three out of the four imidazolyl groups are coordinated to the Pd(II) ion, the fourth coordination site being filled by a chloride anion.

Rh(III) cage compounds have been synthesized by reacting diphenylglycoluril based ligands with RhCl$_3 \cdot$H$_2$O. It can be expected that by this procedure metallo-cage compounds are obtained which contain axially coordinated chloride anions. The chloride anion located inside the cavity would be completely encapsulated and would prevent the cage from collapsing.

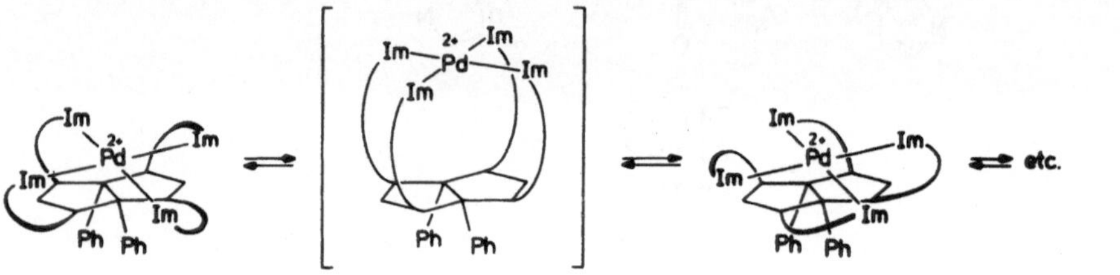

Fig. 14. Process of twisting motion in [Pd(**24**)]$^{2+}$. (Reproduced with permission from the American Chemical Society)

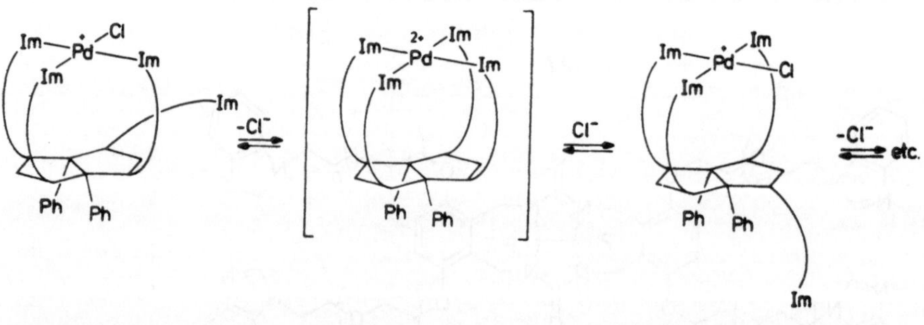

Fig. 15. Fluxional behavior of the Pd(II) complexes derived from **25–27**. (Reproduced with permission from the American Chemical Society)

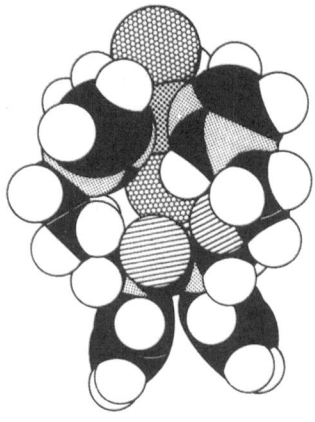

Fig. 16. Model of *trans*-[Rh(**24**)Cl$_2$]$^+$

Starting from ligands **24** and **25** (spacer groups CH$_2$OCH$_2$CH$_2$ and CH$_2$(OCH$_2$CH$_2$)$_2$, respectively) monomeric coordination compounds were formed with conductivities in the range expected for 1:1 electrolytes. The complex with the shorter spacer groups is quite strained because there is very little space for the chloride ligand to fit inside the cavity (Fig. 16). The complex prepared from the ligand with the CH$_2$(OCH$_2$CH$_2$)$_3$ spacers showed broad signals in the ^{1}H-NMR spectrum and multiple resonances for the imidazolyl carbon atoms in the ^{13}C-NMR spectrum. This metallo cage presumably has a structure in which one imidazolyl group coordinates axially to the rhodium center and two chloro ligands coordinate in the *cis* position. Support for this structure comes from a comparison of the far IR spectra of the Rh(III) complexes of **24** and **25** with the model compound trans-[Rh(1-methyl-imidazole)$_4$Cl$_2$]Cl.

Cage compounds with large cavities have been synthesized by using the benzimidazole diphenylglycoluril derivatives **28**. These rigid ligand molecules contain a cleft of approximately 6.5 Å in width (see also Fig. 2), sufficiently large to accomodate a potential guest molecule. The synthesized complexes have structures corresponding to the formula *trans*-[Rh(**28**)Cl$_2$]Cl. The size of the benzimidazole groups prevents axial coordination around the central Rh(III) ion and forces all four ligand arms to adopt positions in the equatorial plane.

4.2 Encapsulated [4Fe-4S] Cluster Complex

Iron-sulfur proteins belong to the class of electron-transport proteins [29]. They contain an iron sulfur cluster, e.g. [4Fe-4S], which shuttles between different oxidation states. The structure of the cluster is quite consistent among a series of these proteins, but their redox potentials vary widely. Synthetic models of iron-sulfur proteins have been designed [30] to investigate the factors that determine the reduction potential of the core and to mimic other biologically important features, such as the cluster environment in a water soluble protein and the creation of a specific iron-subsite.

The cavity of diphenylglycoluril derivative **3** is well suited to partially encapsulate a [4Fe-4S] cluster. Compound **29** which contains four arms terminating with thiol groups was synthesized and treated with $\{(n\text{-Bu})_4N\}_2\text{-}(Fe_4S_4Cl_4)$ in dimethylformamide to give cluster complex **30** [31]. The product was characterized by a number of techniques, including cyclic voltammetry and differential pulse polarography. The current response of **30** was very small, but improved upon addition of a modulator, e.g. Ba^{2+} or Na^+ ions. This behavior is similar to that observed for certain redox active enzymes [32]. As in the natural systems, a maximum response is observed when the Ba^{2+} concentration is

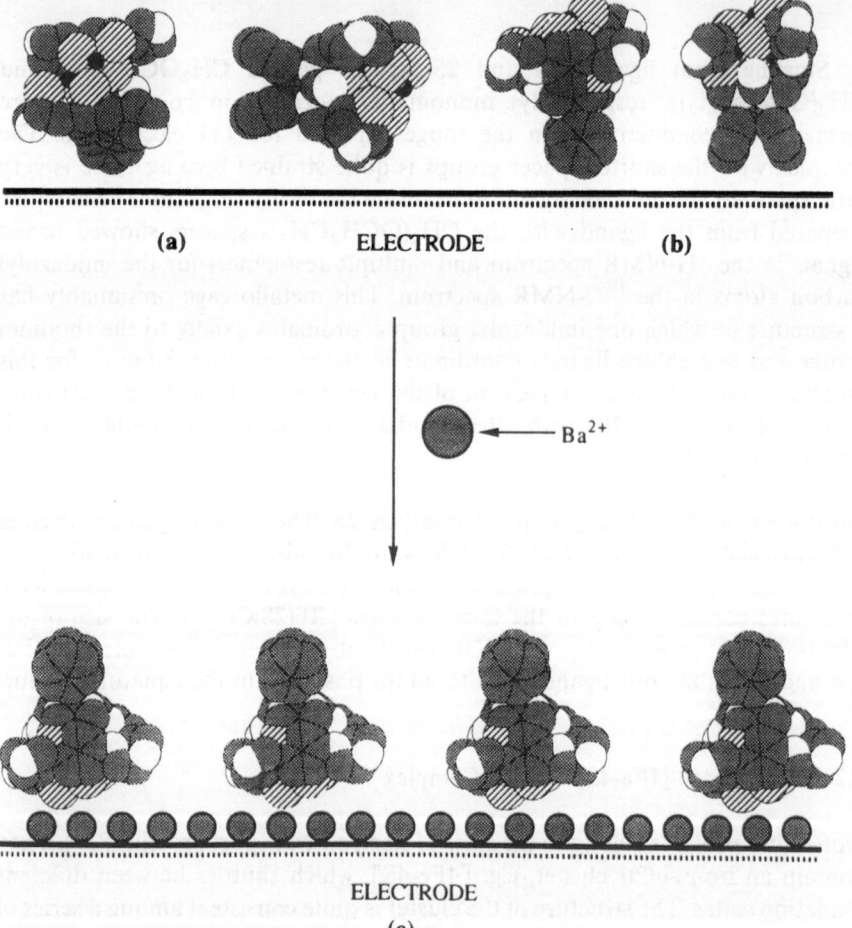

Fig. 17a–c. Possible interactions of **30** with the electrode surface. Sideways docking **a**, backwards docking **b**, and "head first docking" **c**

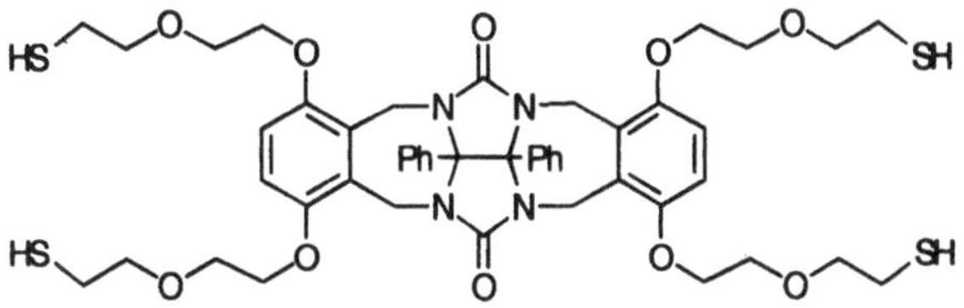

29

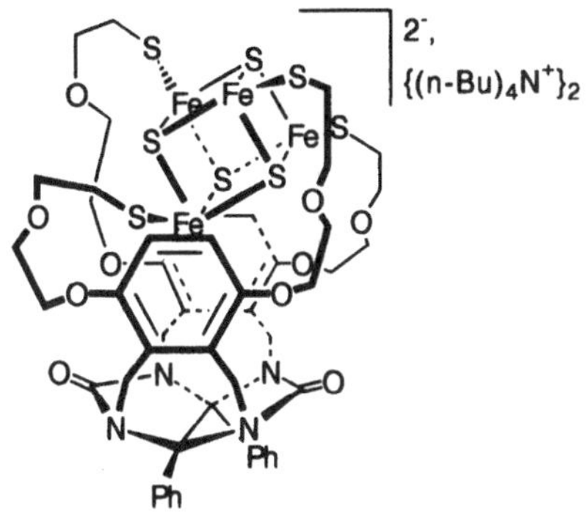

30

approximately 25 mM. It is believed that without any modulator the cluster complex **30** is oriented sideways or backwards with respect to the electrode surface. As a result electron transfer is hampered (see Fig. 17a and b). In the presence of the modulator the complex changes orientation and electron transfer becomes possible (Fig. 17c).

5 Catalysis

5.1 Manganese(III) Catalyst

Recently, diphenylglycoluril-based host compounds have been used to construct catalysts that mimic certain aspects of enzymatic catalysis, e.g. rate enhancement and selectivity. The goal is to bring a binding site and a catalytic center in close proximity.

With the objective of developing a selective epoxidation catalyst, molecular clip **3a** was functionalized with Ni(II)- or Mn(III)-salophen groups (see **31a** and **31b**) [33]. Salen and salophen complexes of Mn(III) are known to epoxidize alkenes in the presence of an oxygen donor [34]. The X-ray structure of the

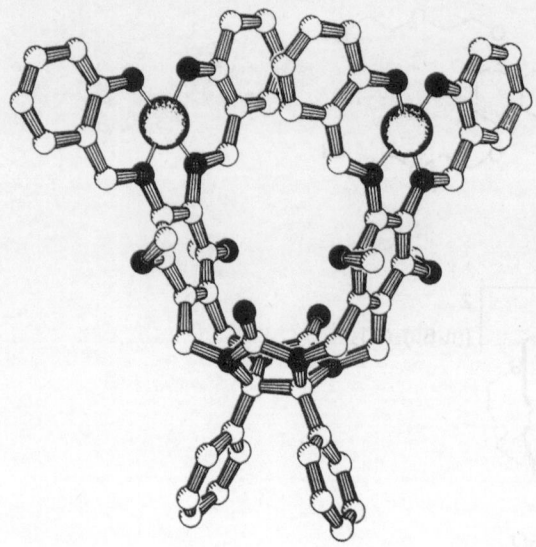

Fig. 18. X-ray structure of **31a**. (Reproduced with permission from the Royal Netherlands Chemical Society)

Ni(II) complex (Fig. 18) reveals that the metalloclip has a twisted structure. Two of the methoxy groups are pointing into the cavity, in this way blocking the access of a guest molecule. As a consequence, resorcinol is not bound in the cleft of **31a**.

The Mn(III) complex **31b** was tested as a catalyst for the epoxidation of various alkenes using sodium hypochlorite or iodosylbenzene as oxidants. Although oxidation took place, no selectivity was observed. For example, allylresorcinol was not epoxidized with rates higher than that of allylbenzene. Presumably, the substrate is not bound in the cleft of **31b** because the latter is occluded by methoxy groups. It is possible that the reaction occurs on the outside of the metalloclip, which cannot discriminate between guest molecules.

31
a, M = Ni(II)
b, M = Mn(III)OAc

50

5.2 Rhodium(I) Catalysts

Rhodium complexes are well-known hydrogenation, hydroformylation, and isomerization catalysts [35]. A supramolecular catalyst containing a tetrakis(triphenylphosphite) rhodium(I) hydride complex linked to a basket molecule (**33**) was synthesized as depicted in Scheme 4 [36]. The synthetic intermediate **32** consists of a mixture of isomers, because the β-diketonate ligand can adopt four different orientations. In one of these isomers, this ligand is located in the cavity of the metallo-host. If $R = Ph$, this conformation does not occur, because the two phenyl groups cannot fit into the cleft. Allylcatechol and allylresorcinol are bound in the cavity of host **16** (this is **33** without the catalyst part) with K_a values of $200\ M^{-1}$ and $3000\ M^{-1}$, respectively. These binding constants are similar to those of catechol and resorcinol in this host. Allyldimethoxybenzene, on the other hand, is not bound in the cavity of **16**. Under 1 atm of hydrogen, the three above mentioned allylbenzene derivatives are converted by catalyst **33** to the corresponding propylbenzene derivatives. The rates of the reduction reactions are related to the K_a values of the substrates: allylresorcinol is converted faster than allylcatechol, which in turn is converted more rapidly than allyldimethoxybenzene. The reference complex $[HRh\{P(OPh)_3\}_4]$, which lacks the substrate binding moiety displays a different activity profile (rate of conversion of allyldimethoxybenzene > allylresorcinol = allylcatechol), suggesting that the cavity of **33** is important for the process of selection.

Complex **34**, which is prepared by treating **33** with carbon monoxide (1 atm) catalyzes the isomerization of double bonds. This catalyst also accelerates the reaction of bound substrates. Under identical conditions (reaction time 2 hours) allylcatechol and allylbenzene are isomerized by **34** to the corresponding methylstyrenes in yields amounting to 68% for the former substrate and only 12% for the latter one [37].

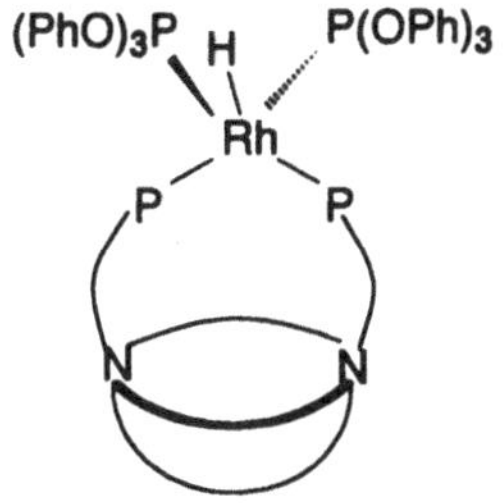

33

5.3 Copper(II) Catalyst

The binding and activation of molecular oxygen is an important process in nature and is achieved with the help of metallo-enzymes, e.g. dicopper enzymes

51

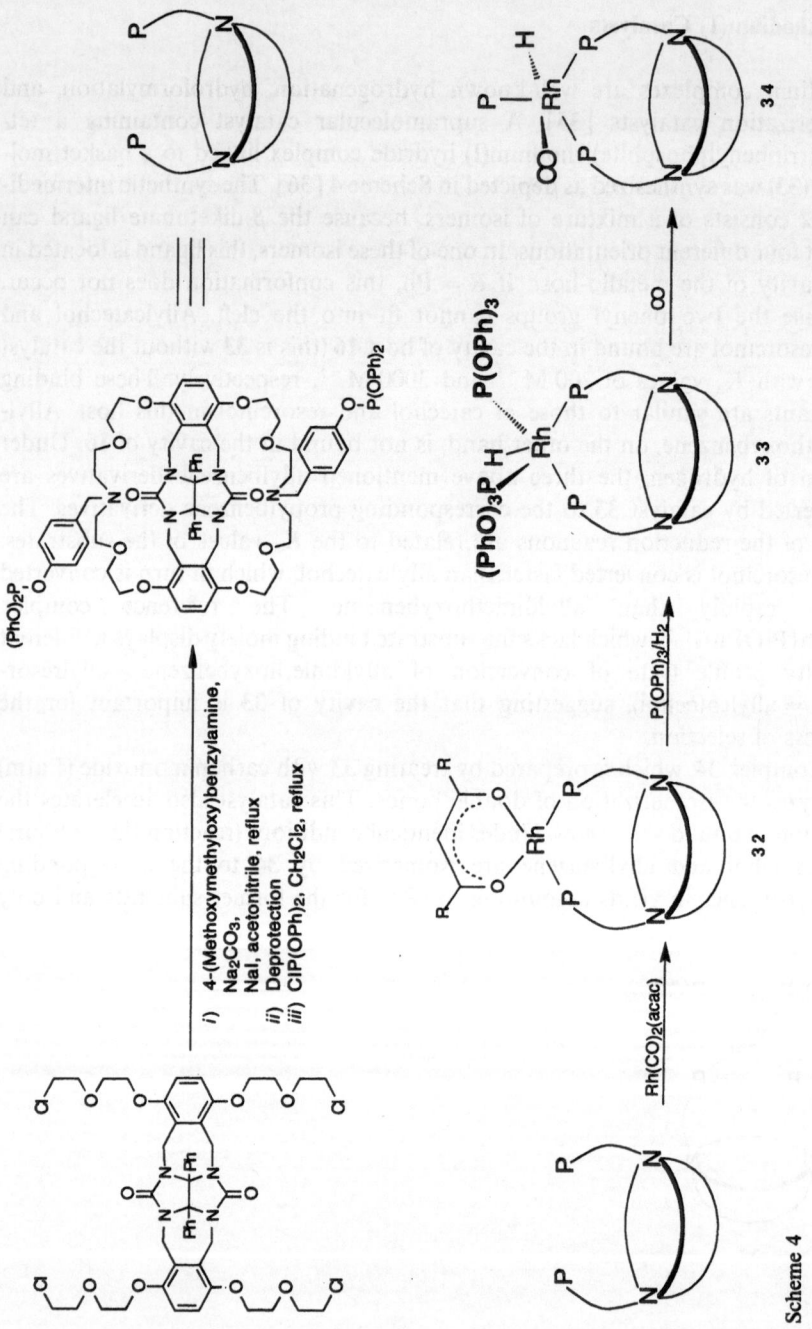

i) 4-(Methoxymethyloxy)benzylamine,
Na$_2$CO$_3$,
NaI, acetonitrile, reflux
ii) Deprotection
iii) ClP(OPh)$_2$, CH$_2$Cl$_2$, reflux

Scheme 4

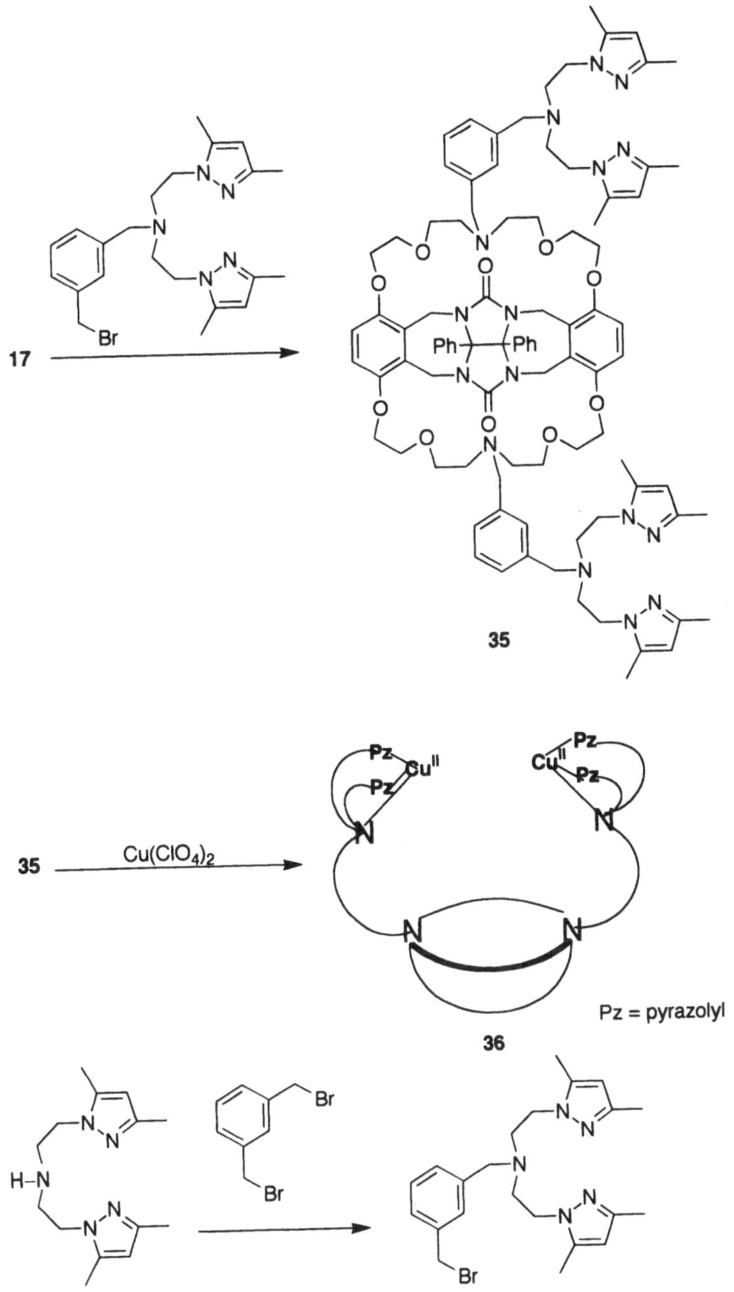

17 $\longrightarrow$

35

35 $\xrightarrow{\text{Cu(ClO}_4)_2}$

36

Pz = pyrazolyl

Scheme 5

such as tyrosinase and dopamine-β-hydroxylase [38]. These enzymes contain a substrate-binding site and two copper centers, which are held by a total of 6 N-donor ligands, usually imidazole groups.

As a first step toward a dopamine-β-hydroxylase mimic, basket-shaped receptor 17 was functionalized with two sets of bis-[2-(3,5-dimethyl-1-pyrazolyl)ethyl]amine ligands to give compound 35 (see Scheme 5). Reaction of 35 with $Cu(ClO_4)_2 \cdot 6H_2O$ yielded complex 36 which was fully characterized [39]. Ligand system 35 binds dopamine derivative 37 and phloroglucinol (1,3,5-trihydroxybenzene) with association constants of $K_a = 60$ and $3500 \, M^{-1}$, respectively. The affinity of the Cu(I) analogon of 36 for 37 was very similar to that of 36 itself and amounted to $K_a = 60 \, M^{-1}$. Resorcinol is bound in the cavity of the Cu(I) complex with a K_a-value of $2000 \, M^{-1}$.

Copper(II) complex 36 is easily reduced to the copper(I) state when an alcohol is present. During this reaction the latter compound is oxidized to an aldehyde. For a series of benzylic alcohols this redox process was studied by UV-vis spectroscopy [39]. The results are presented in Fig. 19 in the form of a Hammett plot. The effect of substituents in the benzene ring on the rate of

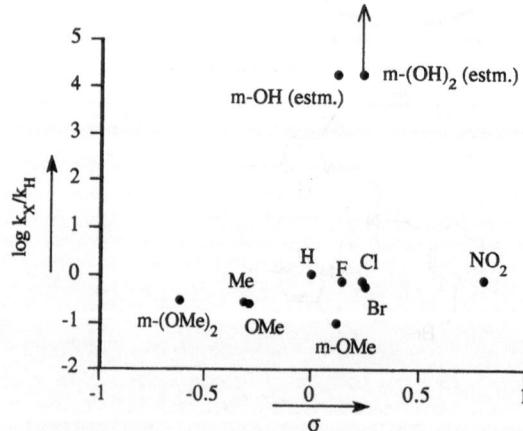

Fig. 19. Hammet plot for the oxidation of benzylic alcohols (X-$C_6H_4CH_2OH$ or X,Y-$C_6H_3CH_2OH$) by complex 36

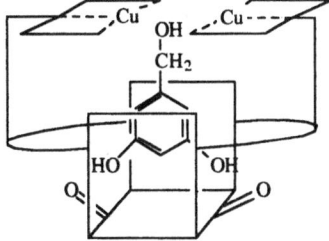

Fig. 20. Orientation of 3,5-dihydroxybenzyl alcohol in the cavity of **36**

benzyl alcohol oxidation is small, except when these substituents are 3-hydroxy or 3,5-dihydroxy functions. In the latter cases rate enhancements are observed equal to or larger than 50 000. These data suggest that the cavity of **36** is capable of selecting substrates with the result that a bound alcohol is oxidized more rapidly than a non-bound one. CPK-models and spectroscopic studies indicate that the dihydroxy-substituted substrate molecule is oriented by the receptor in such a way that its alcohol function is positioned exactly between the two copper centers (Fig. 20). This leads to the very fast reaction that is observed.

6 References

1. Diederich F (1991) Cyclophanes. Royal Society of Chemistry, Cambridge
2. (a) Gutsche CD (1987) Progress in Macrocyclic Chemistry 3: 93; (b) Gutsche CD (1989) Calixarenes. Royal Society of Chemistry, Cambridge
3. (a) Collet A (1987) Tetrahedron 43: 5725 (b) Collet A, Dutasta J-P, Lozach, B (1990) Bull Soc Chim Belg 99: 617
4. Högberg AGS (1980) J Org Chem 45: 4498
5. Rebek J Jr (1990) Angew Chem 102: 261; (1990) Angew. Chem. Int. Ed. Engl. 29: 245
6. Webb TH, Wilcox CS (1990) J Org Chem 55: 363
7. Bender ML, Komiyama M (1978) Cyclodextrin Chemistry. Springer, Berlin
8. For a review of cucurbituril chemistry, see the contribution by W.L. Mock in this volume of Topics in Current Chemistry
9. Smeets JWH, Sijbesma RP, Niele FGM, Spek AL, Smeets WJJ, Nolte RJM (1987) J Am Chem Soc 109: 928
10. Sijbesma RP, Nolte RJM (1993) Rec Trav Chim Pays-Bas 112: 643
11. (a) Sijbesma RP, Kentgens APM, Nolte RJM (1991) J Org Chem 56: 3199; (b) Sijbesma RP, Kentgens APM, Lutz ETG van der Maas JH, Nolte RJM (1993) J Am Chem Soc 115: 899
12. Schuurmans R, (1994) Thesis, Nijmegen
13. For a discussion of hydrogen bonds with the π electrons of carbonyl groups, see: (a) Laurence C, Berthelot M, Hilbert M (1985) Spectrochim Acta 41A: 883; (b) Massat A, Guillaume Ph, Doucet JP, Dubois JE (1991) J Mol Struct 244: 69
14. Johnson CS Jr, Bovey FA (1958) J Chem Phys 29: 1012
15. Sijbesma RP, Bosman WP, Nolte RJM (1991) J Chem Soc Chem Commun 885
16. Bosman WP, Beurskens PT, Admiraal G, Sijbesma RP, Nolte RJM (1991) Zeitschr Kristallogr 197: 305
17. Sijbesma RP, Wijmenga SS, Nolte RJM (1992) J Am Chem Soc 114: 9807
18. Koshland Jr, DE (1958) Proc Natl Acad Sci US 44: 98
19. Smeets JWH, Sijbesma RP, van Dalen I, Spek AL, Smeets WJJ, Nolte RJM (1989) J Org Chem 54: 3710

20. Sijbesma RP, Nolte RJM (1991) J Org Chem 56: 3122
21. Schuurmans R, Scheeren JW, Nolte RJM, Unpublished results.
22. Smeets JWH, Visser HC, Kaats-Richters VEM, Nolte RJM (1990) Rec Trav Chim Pays Bas 109: 147
23. Smeets JWH, van Dalen L, Kaats-Richters VEM, Nolte RJM (1990) J Org Chem 55: 454
24. Martens CF, Sijbesma RP, Klein Gebbink RJM, Spek AL, Nolte RJM (1993) Rec Trav Chim Pays Bas 112: 400
25. Sijbesma RP, Nolte RJM (1992) J Phys Org Chem 5: 649
26. Sijbesma RP, Nolte RJM (1991) J Am Chem Soc 113: 6695
27. Niele FGM, Nolte RJM (1988) J Am Chem Soc 110: 172
28. Niele FGM, Martens CF, Nolte RJM (1989) J Am Chem Soc 111: 2078
29. Lovenberg W ed. (1973) Iron-Sulfur Proteins. Academic Press, New York vol I and II
30. (a) Kuroda Y, Sasaki Y, Shiroiwa Y, Tabushi I (1988) J Am Chem Soc 110: 4049; (b) Okuno Y, Uoto K, Tomohiro T, Youinou M-T (1990) J Chem Soc Dalton Trans 3375; (c) Stack TDP, Holm RH (1988) J Am Chem Soc 110: 2484
31. Martens CF, Blonk HL, Bongers T, van der Linden JGM, Beurskens G, Beurskens PT, Smits JMM, Nolte RJM (1991) J Chem Soc Chem Commun 1623
32. Armstrong FA, Hill HAO (1988) Acc Chem Res 21: 407
33. Gosling PA, Sijbesma RP, Spek AL, Nolte RJM (1993) Rec Trav Chim Pays Bas 112: 404
34. (a) Srinivasan K, Michaud P, Kochi JK (1986) J Am Chem Soc 108: 2309; (b) Yoon H, Burrows CJ (1988) J Am Chem Soc 110: 4087
35. James BR (1973) Homogeneous Hydrogenation, Wiley, New York
36. Coolen HKAC, van Leeuwen PWNM, Nolte RJM (1992) Angew Chem 104: 906; (1992) Angew Chem Int Ed Engl. 31: 905
37. Coolen HKAC (1994) Thesis, Nijmegen
38. Karlin KD, Gultneh Y (1987) Prog Inorg Chem 35: 219
39. Martens CF, Klein Gebbink RJM, Feiters MC, Nolte RJM (1994) J Am Chem Soc 116: 5667

Concave Acids and Bases

Ulrich Lüning

Institut für Organische Chemie, Universität Kiel, Olshausenstr. 40/60, 24098 Kiel, Germany

Table of Contents

Topics in Current Chemistry, Vol. 175
© Springer-Verlag Berlin Heidelberg 1995

The selectivity of an enzyme is largely determined by the geometry of the active site which is usually embedded in a concave region of the protein (cavity or cleft). Similar geometries for standard reagents of organic chemistry can be generated when functional groups like pyridine, 1,10-phenanthroline or benzoic acid are placed into concave structures, e.g. bimacrocycles. In this review, the concept of concave reagents and catalysts, their syntheses and characterization, and their behavior in model reactions will be presented. The bimacrocyclic concave reagents may be used as acids in protonations and as bases in base catalyses. To facilitate the recovery of the concave acids and bases, their fixation to polymers is also described.

1 Introduction

Inside the cell, numerous chemical processes take place at the same time. The cell solves the problem of generating a large number of different molecules at the same time with a high selectivity by the use of enzymes. The selectivity of these proteins is largely determined by their geometry. Also the selectivity of another class of proteins, the receptors, is influenced by geometrical features. Receptors and enzymes have in common that they are equipped with concave structures such as clefts or cavities [1] in which substrate molecules are bound or chemically modified.

Supramolecular Chemistry tries to mimic these concepts, and numerous host molecules have been synthesized [2, 3]. For the understanding of host-guest systems, first the binding of a guest within a host was studied. Numerous binding constants have been measured for inorganic and organic guests in various hosts (H) [2–4]. But these association constants only reflect the thermodynamics of the first step of the reaction, the free binding enthalpy for the educt (E), $\Delta G_{E \cdot H}$. For the understanding of a reaction or a catalysis, kinetic data must be known. The kinetic of a reaction in a host-guest system (i.e. also an enzyme-substrate system) is determined by three rates: the rate of binding of the educt E ($k_{E \cdot H}$), the rate of reaction of the bound substrate E·H to a bound product P·H via the transition state TS·H ($k_{TS \cdot H}$), and the release rate of the product P from the host H (k_P). The rate of the reaction sequence is determined by the slowest of these three rates. For example: If the product releasing step (k_P) is slow, a product inhibition is occuring, limiting the turn-over. The rate constants $k_{E \cdot H}$, $k_{TS \cdot H}$, and k_P are correlated to the enthalpies of activation $\Delta G^{\#}_{E \cdot H}$, $\Delta G^{\#}_{TS \cdot H}$, and $\Delta G^{\#}_P$. Fig. 1 shows the reaction of an educt E to a product P in the absence and in the presence of a host.

If the rate of complexation of the educt E ($k_{E \cdot H}$) and the rate of decomplexation of product P (k_P) are fast, the overall rate of the catalysis of the reaction E → P is determined by the enthalpy of activation $\Delta G^{\#}_{TS \cdot H}$, the enthalpy difference between the bound transition state TS·H and the bound educt E·H. Only when $\Delta G^{\#}_{TS \cdot H}$ is smaller than the enthalpy difference between the free transition state TS and the non-bound educt E ($\Delta G^{\#}_{TS}$), a rate enhancement by catalysis can be observed [5]. Although binding of an educt E and the transition state TS leads to lower free enthalpies for E·H and TS·H, the difference, the activation enthalpy $\Delta G^{\#}_{TS \cdot H}$ need not be smaller. In contrast, if the host H is very selective towards the educt E and binds the educt E better than the transition state TS, the enthalpy of activation $\Delta G^{\#}_{TS \cdot H}$ ($= G_{TS \cdot H} - G_{E \cdot H}$) will increase. This effect may be called "educt inhibition" (see case C in Fig. 1).

Therefore in this review, an approach to reagents is described which are selective due to their concave geometry but do not need binding of the substrate. A standard reagent of organic chemistry with known reactivity and selectivity is chosen and incorporated into a concave environment giving *concave catalysts* and *concave reagents*. Their geometry can be compared with a lamp: the lamp shade is the concave shielding, the light bulb the reactive functionality. Binding is not necessary, the differences in accessibility of the functional group for different substrates or different orientations of a substrate will determine the selectivity which should differ from the selectivity of the non-concave parent reagents.

In contrast, in most traditional host-guest complexes binding occurs first. Then reactions may take place. They can be intermolecular – the complex reacts with a third compound in solution [6] – or intramolecular – a functional group in the host reacts with the guest. But usually, the functional group is just attached to the host. It has no defined position.

On the other hand, in concave reagents, the reactive group is located within the cavity [7]. The concave geometry of enzymes is translated into artificial reagents. The concave shape is retained but the reactive group in the active site is replaced by a standard reagent of organic chemistry. Furthermore the concave

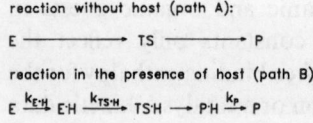

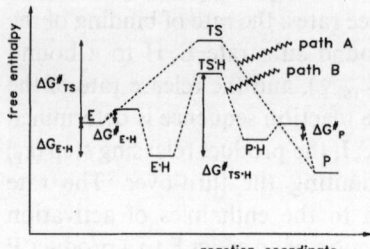

Case A: Educt E and transition state TS are both stabilized upon complexation by the host H. Therefore: $\Delta G^{\neq}_{TS} = \Delta G^{\neq}_{TS \cdot H}$.

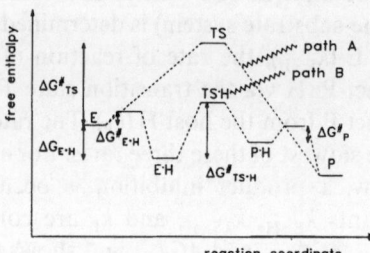

Case B: Upon complexation, the stabilization of the transition state TS is much larger than than the stabilization of the educt E or the product P. The reaction is catalyzed!

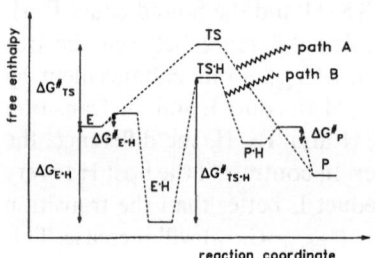

Case C: Upon complexation, the educt E is stabilized most. The rate of reaction is slow, "educt inhibition" occurs.

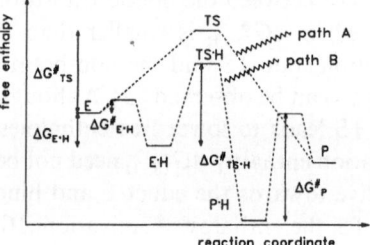

Case D: Upon complexation, the product P is stabilized most. The rate of reaction is slow, "product inhibition" occurs.

Fig. 1. Reaction coordinate/free enthalpy diagram for the reaction of an educt E to a product P in the absence (path A) and in the presence (path B) of a host H. In case (**A**), the binding has no effect on the overall rate while case (**B**) shows catalysis. In case (**C**) "educt inhibition", in case (**D**) "product inhibition" is shown. In cases (**C**) and (**D**), the differences of free enthalpy between E and E·H, and P and P·H, respectively, are much larger than the difference between TS and TS·H

structure need not be made from amino acids. Almost any building block of organic chemistry may be used. The advantages are as follows:

- The structural components are not limited to one group of compounds (amino acids). Therefore the construction of a concave reagent becomes more variable.
- The variation of the functional group in the active site will be easier.
- In comparison to enzymes, the molecular weights will be reduced because large parts of the amino acid framework can be omitted.
- By using groups other than amino acids, concave reagents will be stable under

conditions where enzymes denaturize: high temperatures, concentrated salt solutions, extreme pH, non-aqueous solvents.

In the case of catalytic systems, the tedious and expensive synthesis of a concave catalyst is compensated by its (theoretically) unlimited recyclability. Reagents, in contrast, are used up in a reaction. Therefore, concave reagents will only be attractive when, after the reaction, the used functional groups can be returned into the active original functionality. They must be "rechargeable". This is trivial for acids and bases but in principle should also be realizable for redox reagents.

Which tasks have to be accomplished to obtain a concave reagent?

a) A functional group which acts as a reagent or a catalyst in a standard reaction of organic chemistry must be chosen.
b) This functionality (light bulb) must be incorporated into a framework (lamp shade) which ensures a concave positioning.
c) The reactivity of the functional group in the new concave surrounding must be investigated. Three cases are possible: (i) The concave shielding does not effect the reactivity. (ii) The reactivity is reduced in comparison to the parent reagent but is still high enough to carry out a reaction. (iii) The concave shielding inhibits a reaction. When two different substrates are reacted with the concave reagent and at least for one substrate the reactivity is altered [case (ii) or (iii)], selectivity changes are expected.

In this review, acids and bases [8] (benzoic acids, pyridines and 1,10-phenanthrolines) will be discussed as functional groups (light bulbs). They have been incorporated into bimacrocyclic structures by bis-α- or bis-*ortho*-substitution of the acids or bases: 2,6-disubstituted benzoic acids and pyridines, and 2,9-disubstituted 1,10-phenanthrolines were used. For the construction of the "lamp shade", attention had to be paid to the components. They had to be unreactive in order not to interfere with the reactive "light bulbs". Therefore, for instance amine nitrogen atoms cannot be used as bridgeheads because they may act as bases.

Section 2 discusses the syntheses of different classes of concave acids and bases. Convergent synthetic strategies were chosen for an easy structural variation of the reagents (modular assembly). Section 3 characterizes the concave acids and concave bases and checks whether the acid/base properties of the parent compounds benzoic acid, pyridine and 1,10-phenanthroline are conserved in the bimacrocyclic structures. In Section 4, the influence of the concave shielding on the reactivity and selectivity of the concave reagents is measured in model reactions. In principle, the concave shielding should be able to influence inter- and intramolecular competitions as well as chemoselectivity and (dia)stereoselectivity. If the reagent is chiral, enantioselectivity should also be observable.

U. Lüning

2 Synthesis

For the translation of the structure "light bulb in a lamp shade" into molecular dimensions, one conceivable geometrical form is a bimacrocycle [9]. The rim of the lamp shade is a cycle. In a concave reagent, it must be large enough to let a molecule or a part of a molecule pass through to make contact with the light bulb. Therefore, it must be macrocyclic. This macrocycle must then be spanned by a bridge which carries the functional group, the light bulb. Therefore, at least bimacrocycles must be synthesized.

The bridge must contain the functional group, a base in a case of a concave base. Furthermore, the bridge must ensure the concave positioning of the functional group. Nitrogen atoms are basic, but a normal amine is sp^3-hybridized and the lone pair may invert. Such an inversion is not possible with a sp^2-hybridized nitrogen atom as in pyridine. By a bis-α-substitution (in the case of pyridine by 2,6-disubstitution) the *in*-orientation of the sp^2-lone pair is defined. Also for other functionalities, this bis-α- or bis-*ortho*-substitution of the aromatic ring carrying the functional group is crucial (see below).

The bridgeheads in a bimacrocycle must connect three bridges each. Therefore they must be trifunctional. This is given when trivalent atoms are used and also when trisubstituted groups are incorporated.

2.1 Atoms as Bridgeheads

As stated above, the bridgeheads in bimacrocycles must be at least trivalent. Therefore carbon and nitrogen atoms can be used. But with carbon atoms the orientation of the fourth substituent complicates the situation because stereo-isomers may be formed. Depending on the orientation, *in-in-*, *in-out-* or *out-out*-isomers will be formed [10]. This problem will not occur when trivalent nitrogen atoms are used. However, nitrogen atoms are basic, and therefore the nitrogen bridgeheads will compete with basic functionalities in the lamp shade of a concave base. To circumvent this problem, nitrogen atoms may be incorporated into amide groups. The conjugation between the lone pair of the nitrogen atom and the carbonyl group decreases the basicity of the bridgehead and leaves the functional group as the only basic center.

2.1.1 Concave Pyridine Bislactams

If the functional group is a pyridine unit, 2,6-disubstituted, bimacrocyclic pyridine bislactams must be synthesized. In principle, three classes of such concave pyridines are conceivable (Structures 1).

Structures 1

The amide groups may be connected directly to the pyridine unit (class **1** and **2**), or pyridine ring and amide group may be separated by a spacer (class **3**). In the case of a direct connection, this connection may be via the carbonyl group (class **1**) or the nitrogen atom (class **2**) of the amide function. Representatives of all three classes have been synthesized [11]. However, basicity measurements have shown [11] that only in class **3** has the basicity of the parent compound pyridine been conserved. Therefore, only class **3** in which the amide bridgeheads are separated from the pyridine unit by a methylene spacer has been investigated more thoroughly [12].

The highest flexibility for a variation of the functional group and the chains X and Y (i.e. the size of the rim of the lamp shade) will be realized when the synthesis of **3** is convergent and modular (Scheme 1). Amide bonds can easily be formed in macrocyclizations [13], therefore macrocyclic diamines **7** and diacyl dichlorides **8** had to be prepared. For the synthesis of macrocyclic diamines **7**, also a large number of reactions are known. However, in this case a reduction of a macrocyclic diamide could not be achieved [11]. Therefore, another route was used: the formation of macrocyclic diimines **6** (bis-Schiff bases) followed by NaBH$_4$ reduction to the macrocyclic diamines **7**. This approach has the advantage that for the construction of macrocyclic diimines **6**, the metal ion template effect [14] may be exploited.

A large number of concave pyridines **3** have been synthesized starting from pyridine-2,6-dicarbaldehydes **4**, heteroatom containing α,ω-diamines **5** and diacyl dichlorides **8** (Scheme 1). In the first macrocyclization, the dialdehyde **4** and the diamine **5** were condensed in the presence of an alkaline earth metal ion to give a complex of a macrocyclic diimine **6** which was then reduced to the macrocyclic diamine **7**. The yields of this reaction are excellent (> 90%) when the size of the metal ion is adjusted to the size of the macrocycle formed. Mg^{2+} was used for the formation of 15-membered, Ca^{2+} for 18-membered and Sr^{2+} for 21-membered rings. The macrocyclic diamines **7** were oils which could be purified in some cases. However, for the synthesis of the concave pyridines **3** the purity of the crude diamines **7** was sufficient.

In the second macrocyclization, the reaction of the diamine **7** with a diacyl dichloride **8**, the high dilution technique was used. Depending on the diamine/diacyl dichloride combination, the bimacrocycles **3** were obtained in yields

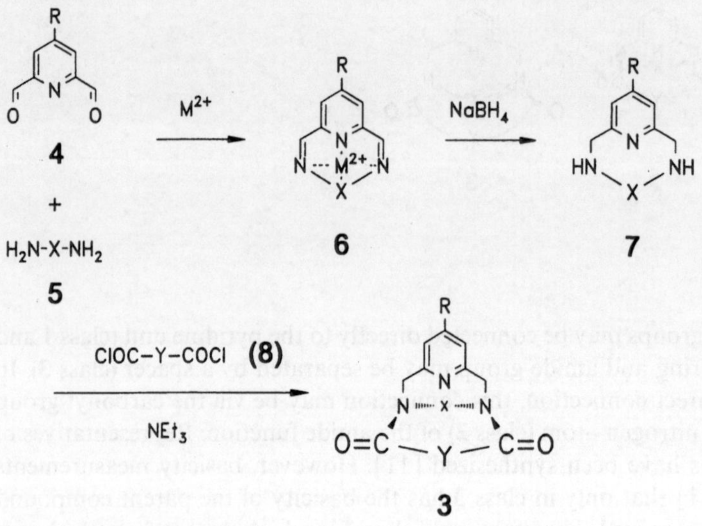

$M^{2+} = Mg^{2+}, Ca^{2+}, Sr^{2+}$

X = polyether

Y = polyether, polymethylene

R = H, OMe, O(CH$_2$)$_2$OH, O(CH$_2$)$_2$OCH$_2$Ph, NEt$_2$

Scheme 1

Table 1. List of known concave pyridine bislactams **3**

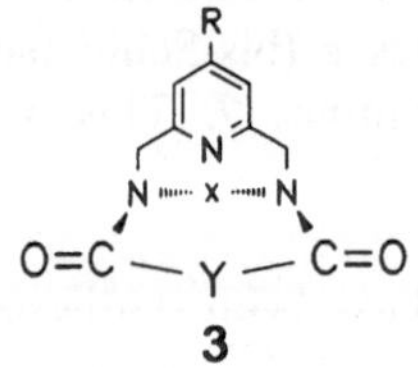

No.	R	X	Y	Ref.
a	H	CH$_2$(CH$_2$OCH$_2$)$_2$CH$_2$	(CH$_2$)$_7$	12b
b	H	CH$_2$(CH$_2$OCH$_2$)$_3$CH$_2$	(CH$_2$)$_7$	15
c	H	CH$_2$(CH$_2$OCH$_2$)$_2$CH$_2$	(CH$_2$)$_{10}$	11
d	H	CH$_2$(CH$_2$OCH$_2$)$_3$CH$_2$	(CH$_2$)$_{10}$	12b
e	H	CH$_2$CH$_2$(CH$_2$OCH$_2$)$_3$CH$_2$CH$_2$	(CH$_2$)$_{10}$	12b
f	H	(CH$_2$)$_3$O(CH$_2$)$_4$O(CH$_2$)$_3$	(CH$_2$)$_{10}$	12b
g	H	CH$_2$(CH$_2$OCH$_2$)$_2$CH$_2$	(CH$_2$OCH$_2$)$_2$	11
h	H	CH$_2$(CH$_2$OCH$_2$)$_2$CH$_2$	(CH$_2$OCH$_2$)$_3$	12b
i	H	CH$_2$(CH$_2$OCH$_2$)$_3$CH$_2$	(CH$_2$OCH$_2$)$_3$	12b
j	OMe	CH$_2$(CH$_2$OCH$_2$)$_2$CH$_2$	(CH$_2$)$_{10}$	12b
k	OMe	CH$_2$(CH$_2$OCH$_2$)$_3$CH$_2$	(CH$_2$)$_{10}$	12b
l	O(CH$_2$)$_2$OH	CH$_2$(CH$_2$OCH$_2$)$_2$CH$_2$	(CH$_2$)$_{10}$	16
m	O(CH$_2$)$_2$OH	CH$_2$(CH$_2$OCH$_2$)$_3$CH$_2$	(CH$_2$)$_{10}$	17
n	O(CH$_2$)$_2$OCH$_2$Ph	CH$_2$(CH$_2$OCH$_2$)$_2$CH$_2$	(CH$_2$)$_{10}$	16
o	O(CH$_2$)$_2$OCH$_2$Ph	CH$_2$(CH$_2$OCH$_2$)$_3$CH$_2$	(CH$_2$)$_{10}$	17
q	NEt$_2$	CH$_2$(CH$_2$OCH$_2$)$_2$CH$_2$	(CH$_2$)$_{10}$	15
r	NEt$_2$	CH$_2$(CH$_2$OCH$_2$)$_3$CH$_2$	(CH$_2$)$_{10}$	12a

up to 61%; the concave pyridines **3** could be synthesized in multigram quantities.

In Table 1, the known [11, 12, 15–17] concave pyridine bislactams are listed.

2.1.2 Concave 1,10-Phenanthroline Bislactams

The synthetic strategy used for the construction of concave pyridine bislactams **3** (Scheme 1) can also be applied to other concave bases. When instead of a pyridine-2,6-dialdehyde **4**, 1,10-phenanthroline-2,9-dicarbaldehyde (**9**) was used in a metal ion template directed synthesis of macrocyclic diimines, after reduction, also macrocyclic 1,10-phenanthroline diamines **10** could be obtained in good yields. Here too, the crude diamines **10** were used in the next reaction step. Bridging of **10** with diacyl dichlorides **8** gave concave 1,10-phenanthroline bislactams **11**. Scheme 2 summarizes the synthesis and lists the synthesized bimacrocycles **11** [18].

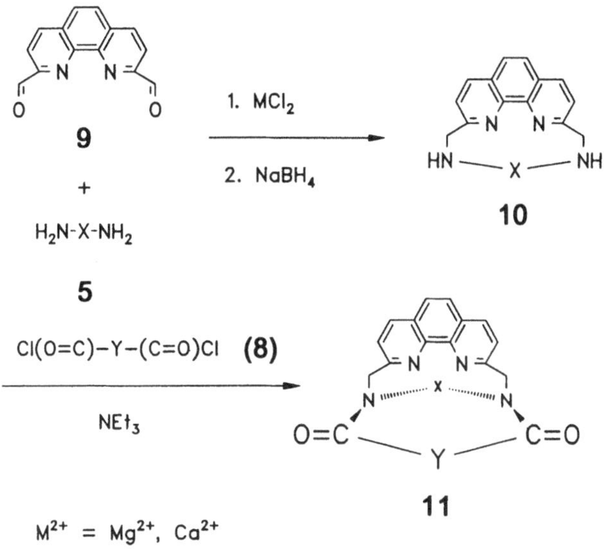

Scheme 2

2.1.3 Concave Pyridine Bissulfonamides

As discussed in Sect. 3, the structure of concave pyridines **3** (Structures 1) is not uniform. Conformers are present due to the hindered rotation in an amide bond.

Therefore, in a second class of concave pyridines 13 (Scheme 3), the carboxamide groups of 3 were replaced by sulfonamides because sulfonamides have a lower barrier to rotation between the nitrogen atom and the sulfonyl group at room temperature [19]. The synthetic strategy for 13 was analogous to the synthesis of the bislactames 3 (s. Scheme 1). Macrocyclic diamines 7 were synthesized by the metal ion template method and then bridged with disulfonyl dichlorides 12. Because sulfonyl chlorides are less reactive towards amines than carbonyl chlorides, this cyclization was catalyzed by 4-dimethylaminopyridine (DMAP) and carried out at 40 °C. The yields of the sulfonamides 13 were lower (max. 30%) but the non-existence of conformers in the bimacrocyclic sulfonamides (see Sect. 3.2) made the isolation easier. Therefore also in this case, gram quantities of 13 could be synthesized easily. In Scheme 3, all known [12a, 15] concave pyridine bissulfonamides 13 are listed.

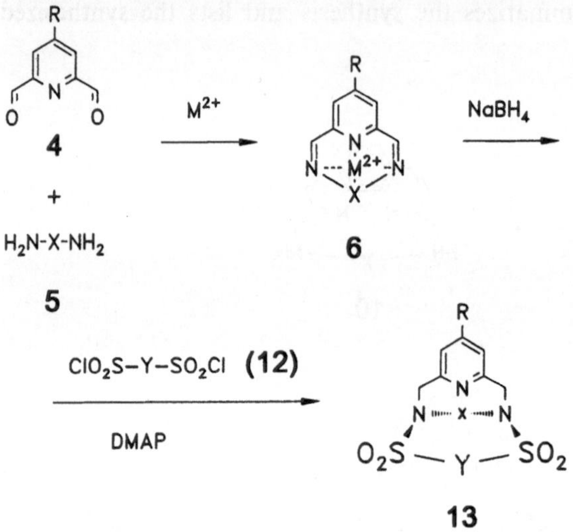

$M^{2+} = Mg^{2+},\ Ca^{2+},\ Sr^{2+}$

No.	R	X	Y	Ref.
a	H	$CH_2(CH_2OCH_2)_2CH_2$	$(CH_2)_{10}$	12a
b	H	$CH_2(CH_2OCH_2)_3CH_2$	$(CH_2)_{10}$	12a
c	H	$CH_2(CH_2OCH_2)_2CH_2$	$(CH_2OCH_2)_3$	12a
d	H	$CH_2(CH_2OCH_2)_3CH_2$	$(CH_2OCH_2)_3$	12a
e	OMe	$CH_2(CH_2OCH_2)_2CH_2$	$(CH_2)_{10}$	12a
f	OMe	$CH_2(CH_2OCH_2)_3CH_2$	$(CH_2)_{10}$	12a
g	OMe	$CH_2(CH_2OCH_2)_2CH_2$	$(CH_2OCH_2)_3$	12a
h	OMe	$CH_2(CH_2OCH_2)_3CH_2$	$(CH_2OCH_2)_3$	12a
i	OMe	$CH_2(CH_2OCH_2)_4CH_2$	$(CH_2OCH_2)_3$	12a
j	NEt_2	$CH_2(CH_2OCH_2)_2CH_2$	$(CH_2)_{10}$	15
k	NEt_2	$CH_2(CH_2OCH_2)_3CH_2$	$(CH_2)_{10}$	12a
l	NEt_2	$CH_2(CH_2OCH_2)_3CH_2$	$(CH_2OCH_2)_3$	12a
m	NEt_2	$CH_2(CH_2OCH_2)_4CH_2$	$(CH_2)_{10}$	12a

Scheme 3

The strategy for the synthesis of concave pyridine bissulfonamides **13** was also applied to the synthesis of concave 1,10-phenanthroline bissulfonamides **14** (Structures 2) but here the yields were disappointing (< 10%) [20]. Therefore, for conformer-free concave 1,10-phenanthrolines the bridgehead atoms have been substituted by bridgehead groups (see below).

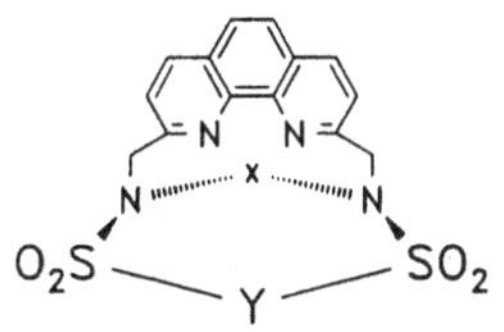

14

	X	Y
a	CH$_2$(CH$_2$OCH$_2$)$_2$CH$_2$	(CH$_2$)$_{10}$
b	CH$_2$(CH$_2$OCH$_2$)$_2$CH$_2$	(CH$_2$OCH$_2$)$_3$

Structures 2

2.2 Groups as Bridgeheads

An alternative to atoms as bridgeheads are trisubstituted groups. To simplify the geometry of the concave reagent, flat and symmetrical structures were used: 1,2,3- and 1,3,5-trisubstituted benzenes. When these groups are incorporated into bimacrocyclic structures cyclophanes [3] are formed. Three different strategies have been used to synthesize concave 1,10-phenanthroline cyclophanes, concave pyridine cyclophanes, and concave benzoic acid cyclophanes. All three methods have in common that the aryl bridgeheads are joined first with the functional group (1,10-phenanthroline, pyridine, or benzoic acid) before a double macrocyclization is used to build up the final bimacrocycles [9].

2.2.1 Concave 1,10-Phenanthroline Cyclophanes

To incorporate an aryl group into the 2- and 9-position of a 1,10-phenanthroline, Sauvage's [21] addition of aryl lithium compounds to 1,10-phenanthroline was used [22] (Scheme 4). In 58% yield, the tetra-*ortho*-methoxy substituted 2,9-diaryl-1,10-phenanthroline **17** was synthesized from 1,10-phenanthroline **15** and the lithium salt **16**. Reaction with BBr$_3$ gave the tetraphenol **18** which could be doubly bridged by a variety of diiodides **19** or ditosylates **20**. Because the tetraphenol **18** has four hydroxy groups, it is also possible that the chains X join two oxygen atoms of the same aryl bridgeheads leading to metacyclophanes **43** (see Sect. 3.1). But only the concave bimacrocycles **21** were isolated. Structures 3 lists the known [20, 22] concave 1,10-phenanthroline cyclophanes **21**. The yields are ≤ 30% but this is acceptable because two macrocycles are formed in one step.

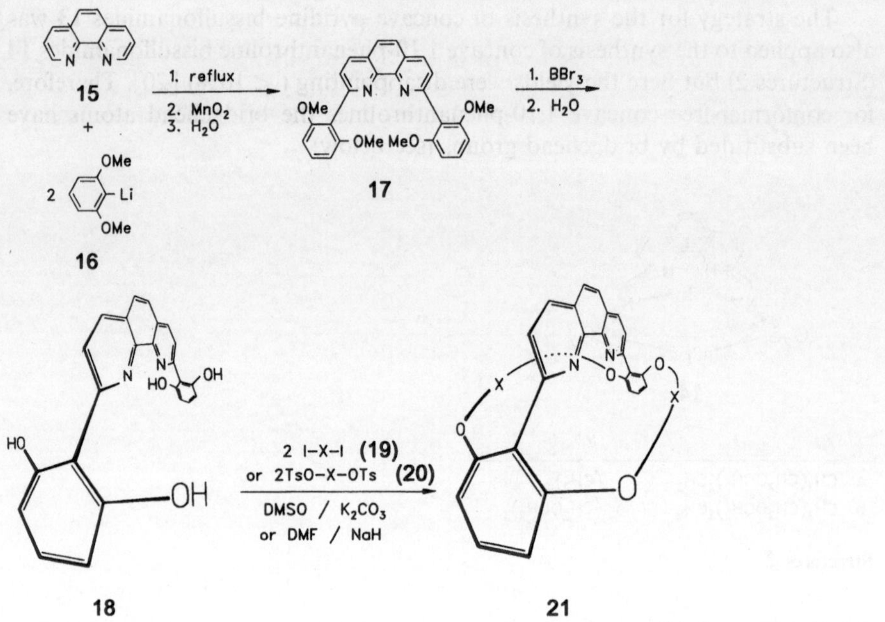
15
+
2
OMe
Li
OMe
16
1. reflux
2. MnO$_2$
3. H$_2$O
OMe
OMe MeO
OMe
17
1. BBr$_3$
2. H$_2$O
HO
HO
OH
OH
18
2 I–X–I (19)
or 2TsO–X–OTs (20)
DMSO / K$_2$CO$_3$
or DMF / NaH
X
X
O
O
21

Scheme 4

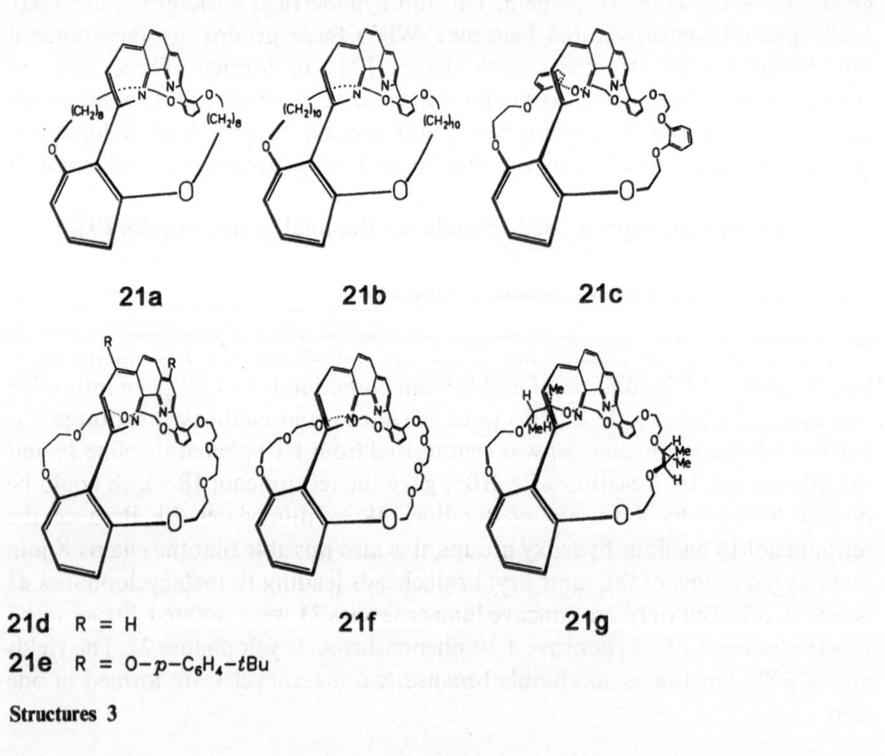
(CH$_2$)$_8$
(CH$_2$)$_8$
21a
(CH$_2$)$_{10}$
(CH$_2$)$_{10}$
21b
21c
R
R
21d R = H
21e R = O–p–C$_6$H$_4$–tBu
21f
Me
H
MeH
Me
Me
H
21g

Structures 3

2.2.2 Concave Pyridine Cyclophanes

In the synthesis of concave 1,10-phenanthroline cyclophanes **21** (Scheme 4), the aryl bridgeheads could be easily introduced by the addition of two aryl lithium moieties **16** to 1,10-phenanthroline (**15**). Although aryl lithium compounds may also be added to pyridine [23], this approach could not be realized for the construction of concave pyridine cyclophanes **29** yet [24]. Therefore another route for the synthesis of the concave pyridines **29** was used (Scheme 5): the cyclization of 1,5-diaryl substituted C_5-units **23** or **27** with ammonia. The resulting 2,6-bis(2,6-dimethoxyphenyl)pyridine **24** [25], the pyridine analogue to the tetramethoxy-1,10-phenanthroline derivative **17**, was then treated in the same way. After liberation of the phenol functions, the four OH groups of **28** were reacted with two equivalents of diiodides **19** [25]. As in the synthesis of the concave 1,10-phenanthroline cyclophanes **21**, two macrocycles were formed in one reaction step.

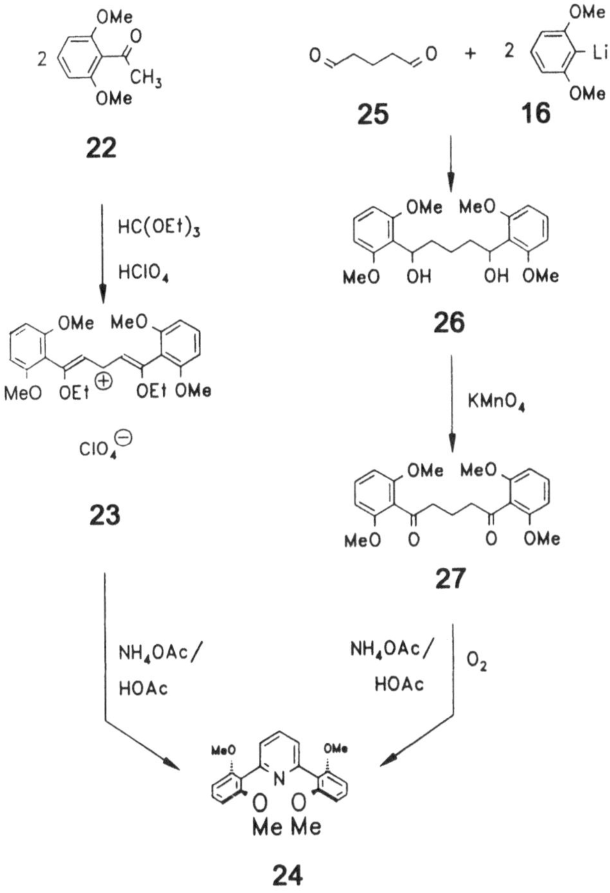

Scheme 5

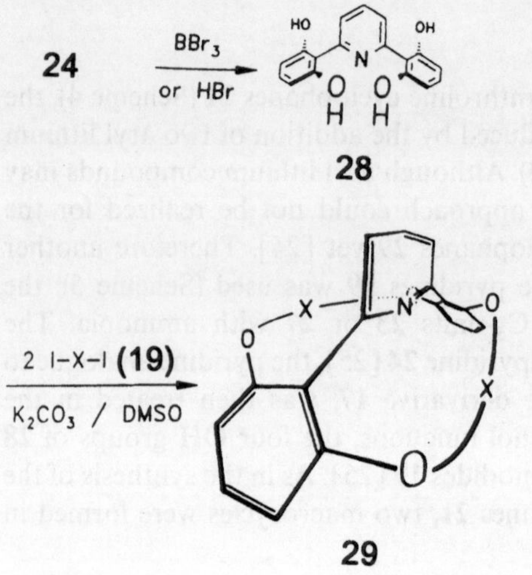

Scheme 5 (contd.)

2.2.3 Concave Benzoic Acid Cyclophanes

If the pyridine ring in a concave pyridine cyclophane **29** was exchanged by a benzoic acid, a concave benzoic acid cyclophane **38** would result (Scheme 6). The core of this acid is a *m*-terphenyl substituted in 2′-position. *m*-Terphenyls can be

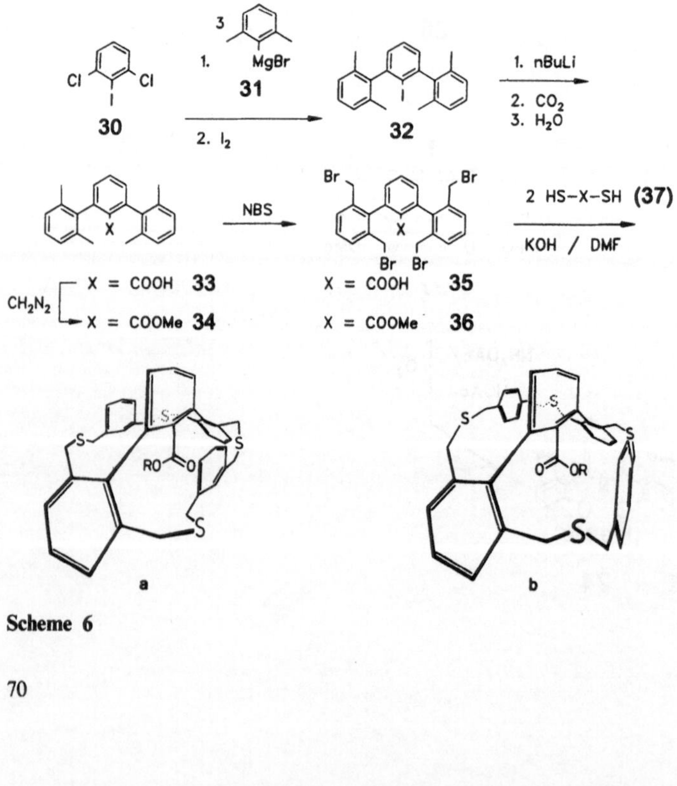

Scheme 6

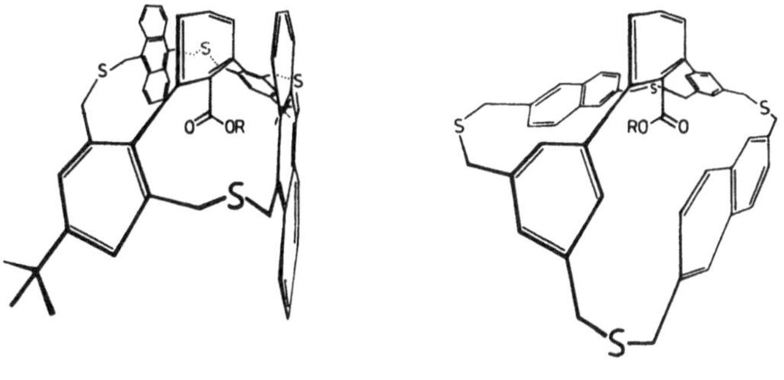

c   d

R = H   **38**

R = Me   **39**

Scheme 6 (contd.)

easily prepared by the reaction of an aryl-Grignard reagent **31** with 1,3-dichloro-2-iodobenzene (**30**) [26]. When the intermediate Grignard functionality in 2'-position is quenched with iodine, the 2'-iodo substituted *m*-terphenyl **32** is generated.

This iodo compound **32** can be lithiated with *n*-butyl lithium to give an organolithium compound. Its reaction with CO_2 gave the *m*-terphenylcarboxylic acid **33** which forms the ester **34** upon treatment with diazomethane [27]. The four methyl groups in 2,2'',6,6''-position of **33** and **34** could be functionalized by NBS-bromination [27]. Subsequent double-bridging of the tetrabromides **35** and **36** with dithiols **37** gave concave benzoic acids **38** and esters **39** [27] as shown in Scheme 6. The more aryl rings a concave acid contained the less soluble it was. To increase the solubility, the anthracyl derivatives **38c** and **39c** were therefore substituted by two *tert*-butyl groups [27b].

The concave acids **38** and the esters **39** could be interconverted. Reaction of the acids **38** with diazomethane led to the methyl esters **39**. S_N2 dealkylation of the esters **39** by lithium iodide in pyridine gave the acids **38** [27].

2.3 Other Concave Reagents

The synthetic strategies discussed above are not restricted to pyridines, 1,10-phenanthrolines and benzoic acids. Therefore a large number of other concave reagents may be synthesized in the future.

As an example, starting from *m*-terphenyls with other functionalities in 2'-position, other concave molecules with a *m*-terphenyl core are conceivable. Sulfur substituents have already been introduced into the 2'-position of a *m*-terphenyl, for instance $X = SH, SO_2H$ or SO_2Cl (see Section 6). A protected thiol functionality has already been incorporated into a concave thiol acetate **42** [28] (Scheme 7).

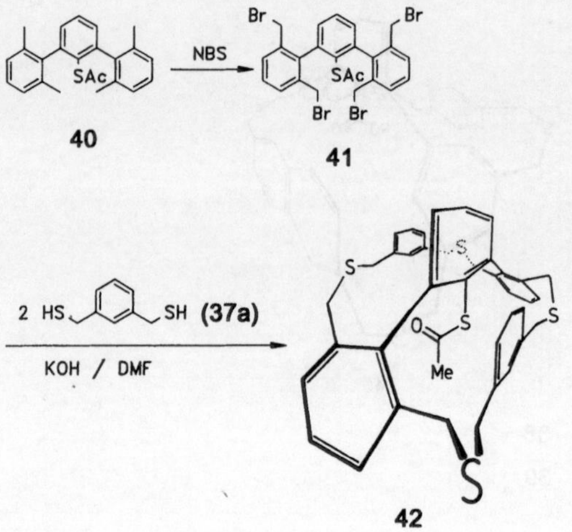

Scheme 7

3 Characterization

All concave acids and bases mentioned here were characterized by standard analytical methods: m.p., elemental analyses, IR, NMR and MS. In some cases the structure was further elucidated by X-ray analysis. Furthermore, the acidity (basicity, respectively) was determined.

3.1 Proof of Assigned Structure

In the synthesis of all concave acids and bases, a difunctionalized molecule A–A was cyclized with a difunctionalized bridge component B–B. Because telo- and polymerizations are the main side reactions [29] the isolated macrocycles need not be the expected [1 + 1] addition products, the (–A–AB–B–)$_1$ cycles. [n + n] Telomers with the general structure (–A–AB–B–)$_n$ are also possible. These molecules have identical elemental analyses and similar IR and NMR data. Therefore the mass spectral analyses of the macrocycles are very important because this is the only method which can tell [1 + 1] and [2 + 2] addition products apart. Due to the high molecular weight of the concave acids and bases, special MS techniques were necessary in some cases [30]. In the case of the macrocyclic diamine 7 [R = NEt$_2$, X = CH$_2$(CH$_2$OCH$_2$)$_2$CH$_2$], a [2 + 2] addition product could be isolated and characterized besides the desired [1 + 1] product [12a].

When aryl rings were used as bridgeheads (Section 2.2), a tetrafunctionalized pyridine 28 (s. Scheme 5), 1,10-phenanthroline 18 (s. Scheme 4), benzoic acid 35 (s. Scheme 6) or methyl benzoate 36 (s. Scheme 6) was reacted with two equivalents of a difunctionalized bridge component **19, 20** (s. Scheme 4 and 5) or **37** (s. Scheme 6). Therefore in all these cases, not only telomers can be formed as undesired side products. Also two isomeric [1 + 2] addition products are conceivable, the desired concave acids (and bases) and bis-cyclophanes in which the bridges have connected the two functional groups of the same aryl ring. When the tetraphenol precursor **18** is bridged with two equivalents of **19** or **20**, two isomers may be formed: the desired bimacrocyclic concave 1,10-phenanthroline **21** and a bis-cyclophane **43** (Structures 4).

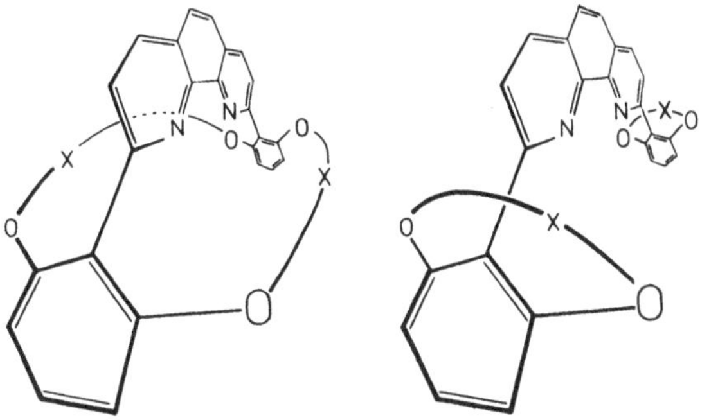

| 21 | 43 |

Structures 4

Whereas the identification of a [1 + 1] or a [2 + 2] cyclization product was unequivocal by MS, it was more difficult to distinguish between the concave bimacrocycles and the bis-cyclophanes. But a combination of NOE investigations with complexation studies followed by NMR, and finally X-ray analyses proved the bimacrocyclic concave structure of the concave acids and bases **21, 29** and **38** [15, 20, 27a].

3.2 Conformational Studies

The first synthesized concave bases, the concave pyridine bislactames **3** (Structures 1), possess two amide groups in each molecule. The rotational barrier for a carboxamide bond is ca. 75 kJ/mol [12b, 19]. Therefore at room temperature, E- and Z-forms are observed in the NMR spectra. Because each concave pyridine bislactame **3** contains two amide groups, diastereoisomeric conformers are observable (see Fig. 2). Structures 5 show the ZZ-, EZ- and EE-conformers for the concave pyridine **3c**.

These conformers do not only complicate the NMR spectra, they also make it difficult to discuss the reactive conformation of the concave reagent. Furthermore, the existence of the conformer mixture slows down the crystallization of

U. Lüning

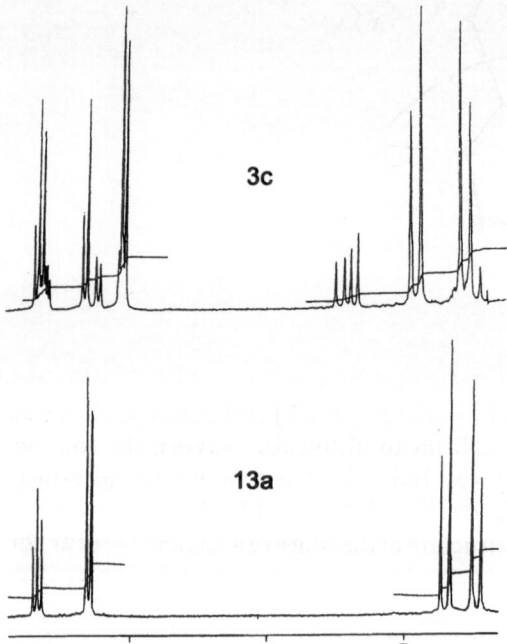

ZZ-**3c**   EZ-**3c**   EE-**3c**

Structures 5

3c

13a

7   6   5

Fig. 2. Comparison of ^{1}H NMR-signals between the concave pyridine bislactame **3c** (Table 1) and the concave pyridine bissulfonamide **13a** (s. Scheme 3). While for **13a** only one set of signals exists for the pyridine H-atoms (> 7 ppm) and the H-atoms of the methylene spacer (< 5 ppm), the spectrum of **3c** shows three sets of signals (ZZ, EZ and EE) [12a]

the compounds enormously. And even in the solid state, the conformers may coexist, as IR spectroscopy and DSC studies have shown [12b]. It was also difficult to obtain crystals suitable for X-ray analysis.

To circumvent these problems, the carboxamide groups have been substituted by sulfonamide groups (see Sect. 2.1.3). Indeed, the concave pyridine bissulfonamides **13** (s. Scheme 3) possessed much simpler NMR spectra. Figure 2 compares the concave pyridine bislactame **3c** with the corresponding concave pyridine bissulfonamide **13a**.

3.3 Basicity, Acidity

A concave base must be concave and basic. Therefore, the basicities (and acidities, respectively) of concave bases and acids have been determined. Because most concave acids and bases are not soluble in water, the measurements had to be performed in an organic solvent. A photometric titration versus thymol blue in ethanol was chosen [31].

As expected, the basicity of concave pyridine bislactams **3** (Structures 1) and pyridine bissulfonamides **13** (s. Scheme 3) was enhanced by 4-substitution with electron-donating groups like OMe or NEt$_2$ (ca. 1 pK_a unit for OMe, more than 3 pK_a units for NEt$_2$ [12, 32]). But also the nature of the chains X and Y had a strong influence on the basicities. The basicities of 4-*H*-substituted pyridines **3** vary by more than two orders of magnitude [12]. However, when the basicities of the concave pyridines **3** are compared with non-concave pyridines they cover the same range of basicity.

It therefore can be stated: The base-property of pyridine is conserved in the concave pyridine bislactams **3**. But the basicity of each individual concave base is determined by the overall structure (X, Y) and the substitution pattern of the pyridine.

When the concave pyridine bislactams **3** are compared to the bissulfonamides **13**, a decrease for the basicity of **13** can be noticed. The exchange of the carbonamide function in **3** by a sulfonamide in **13** lowers the basicity by up to two orders of magnitude [12a].

In addition, the basicities (or acidities) of the concave 1,10-phenanthrolines **11** (s. Scheme 2) and **21** (s. Scheme 4) [18, 20] and the concave benzoic acids **38** (s. Scheme 6) [27, 33] have been measured. Also in these compounds, the acid/base properties of the parent compounds 1,10-phenanthroline and benzoic acid are at least conserved. Surprisingly, in the concave 1,10-phenanthroline cyclophanes **21** distinctly higher basicities than for 2,9-unsubstituted 1,10-phenanthrolines were found (ca. two orders of magnitude). Substitution in 4,7-position by aryloxy-substituents **21e** (s. Structures 3) further increased the basicity [20].

3.4 X-Ray Analyses

Besides the examination of the acid/base properties of the concave acids and bases, the geometry of the new bimacrocyclic acids and bases had to be investigated to show that the concave acids and bases were indeed concave. Therefore crystal structures were elucidated by X-ray analyses. For the pyridine bislactam **3e** (s. Table 1) [12b], the pyridine bissulfonamide **13j** (s. Scheme 3) [15], the 1,10-phenanthroline cyclophane **21a** (Structures 3) [20] and the benzoic acid **38a** (s. Scheme 6) [27a], crystals could be obtained which were suitable for X-ray analyses. The structures are shown as stereoplots in Figs. 3(a)–(d).

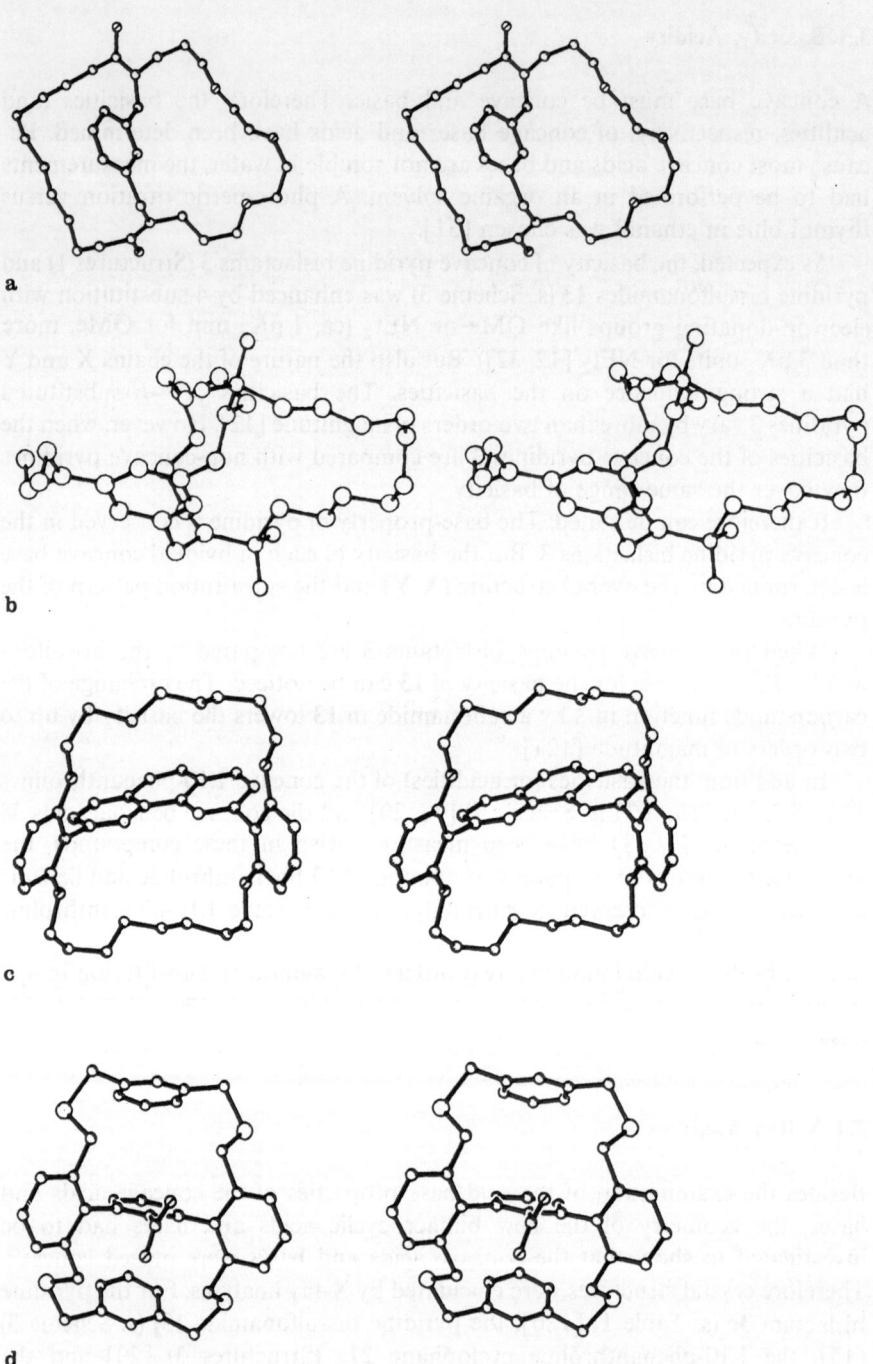

Fig. 3. Stereoplots of the crystal structures of the concave acids and bases: (a) **3e**, (b) **13j**, (c) **21a** and (d) **38a**

Besides a further proof of the bimacrocyclic structures, the X-ray analyses clearly show that the degree of concavity depends on the size and conformation of the bimacrocycles. While the structure of the bislactame **3e** resembles the model "light bulb in a lamp shade", in the bissulfonamide **13j**, the pyridine ring lies flat in one plane with the polymethylene chain with the polyether chain shielding one side. In this conformation, a frontal attack at the sp^2 nitrogen atom of the pyridine ring is shielded by the polymethylene chain. If a reaction takes place at the pyridine nitrogen atom, an attack from below is therefore most probable. But the sp^2 lone pair orbital is oriented orthogonal to a hypothetical attack from below. This may be a reason for the low reactivity of concave pyridine bissulfonamides **13** in comparison to concave pyridine bislactams **3** (see Sect. 4.2) if in solution, the same conformation existed for a concave pyridine bissulfonamide **13** as in this X-ray.

The other three X-ray structures show the expected lamp-like geometry. While for **3e** in solution conformers are known (see Sect. 3.2.) and the chains X and Y may have varying orientations, such flexibilities are not existent in the concave 1,10-phenanthroline **21a** and in the concave benzoic acid **38a**. Here the X-ray structure should be much closer to the conformations in solution.

The data of the X-ray analyses were then used in computer studies to examine the concavity of the compounds more thoroughly.

3.5 Surface Studies

Based on the X-ray data (see above), the accessibility of the concave functional group was studied. By computer modelling (Connolly-routine [34]), spheres of varying sizes were rolled over the van der Waals-surfaces of the concave reagents which were calculated from the X-ray data, and the resulting contact surface was monitored.

In the case of the concave acid **38a** (s. Scheme 6), the Connolly routine showed that the *m*-phenylene chains were not sufficient to ensure a concave environment [33]. Unfortunately no crystals of the concave acids **38b** and **38c** (s. Scheme 6) with more extended chains X could be obtained which were suitable for an X-ray analysis. But NMR studies of the corresponding concave methyl benzoates **39** (s. Scheme 6) showed an increase in shielding by the chains X: the larger the chains X were (**39a** → **39b** → **39c**) the more upfield the methyl signal of the ester groups in **39a–c** was shifted in the ¹H NMR spectra [27b].

When the Connolly-routine was applied to the concave 1,10-phenanthroline **21a** (Structures 3), spheres with a radius up to 2.7 Å were able to touch the nitrogen atoms whereas larger spheres were not. This means that spheres with a diameter of more than 5.5 Å cannot react with the basic nitrogen atoms of **21a** anymore [20].

In the case of the crystallized pyridine **3e** (s. Table 1), for the Connolly-routine a proton was attached to the pyridine atom and the contact between the rolling sphere and this proton was examined. Also here, small spheres may contact the proton whereas large ones may not. In this case, the cut-off radius of the sphere was between 2.0 and 2.5 Å [12b, 32].

The above computer study suggested to investigate protonation reactions via protonated concave pyridines **3**. Size selectivity was expected. In the following section (4.1) intramolecular competitions for protonations by concave pyridines **3** will be discussed.

4 Reactions

4.1 Model Reactions with Concave Acids

4.1.1 The Soft Nef-Reaction

As the Connolly studies suggested (Sect. 3.5), protonated concave pyridines should be able to discriminate between small and large molecules. In a model reaction, a protonation reaction has therefore been examined. The protonation of nitronate ions **44** has been chosen [35] (Scheme 8). In these ions an intramolecular competition of carbon versus oxygen protonation leads to nitro (**45**) or *aci*-nitro (**46**) compounds. The latter ones may then be hydrolyzed by way of the Nef-reaction [36] to form carbonyl compounds **47**.

C–protonation

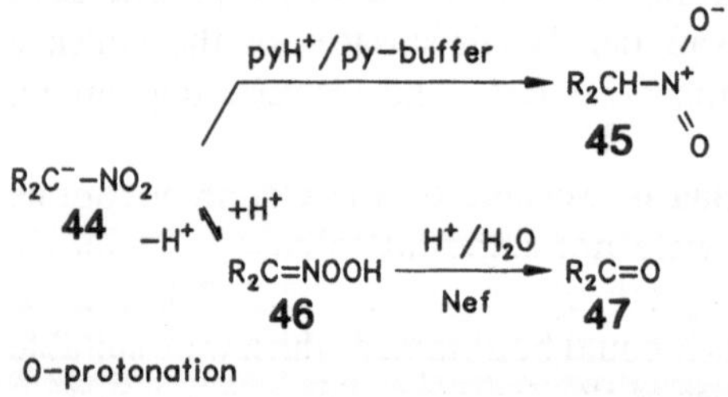

$$R_2C^- - NO_2 \quad \xrightarrow{\text{pyH}^+/\text{py-buffer}} \quad R_2CH-N^+\begin{smallmatrix}O^-\\ \\ \\O\end{smallmatrix}$$

44

45

$$\xrightarrow[-H^+]{+H^+} \quad R_2C=NOOH \xrightarrow[\text{Nef}]{H^+/H_2O} R_2C=O$$

46   **47**

O–protonation

Scheme 8

While the Nef-reaction (i.e. O-protonation) needs strong acidic conditions [36] the formation of nitro compounds (C-protonation) is achieved in buffers, e.g. in pyridinium/pyridine buffers. But when the pyridine of the buffer was exchanged by a concave pyridine **3** (Table 1) of the same basicity while keeping the concentrations constant, the course of the reaction changed completely: no C-protonation was found, the only products were carbonyl compounds **47** although the reaction was carried out in a buffer. Therefore, the concave shielding in the concave pyridine **3** must be responsible for this result because the concave pyridine buffer had the same pH (i.e. identical $EtOH_2^+$ concentration) and the same concentration of protonated pyridine as the pyridine buffer. Scheme 9 explains the situation.

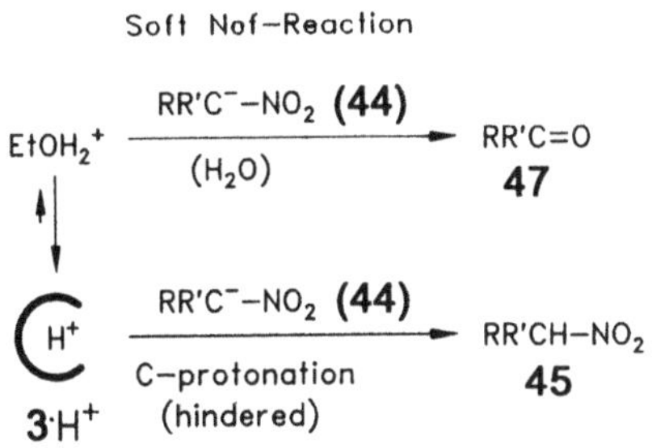

Soft Nef—Reaction

$EtOH_2^+$

$$\xrightarrow[\text{(H}_2\text{O)}]{\text{RR'C}^- - \text{NO}_2 \ (\mathbf{44})} \text{RR'C=O} \quad \mathbf{47}$$

$$\overset{\curvearrowright}{\underset{\mathbf{3 \cdot H^+}}{\left(\underset{\text{H}^+}{}\right.}} \xrightarrow[\substack{\text{C-protonation} \\ \text{(hindered)}}]{\text{RR'C}^- - \text{NO}_2 \ (\mathbf{44})} \text{RR'CH—NO}_2 \quad \mathbf{45}$$

Scheme 9

Due to the basicity of the (concave) pyridine, in a buffer most protons are located on the pyridine nitrogen. The concentration of free $EtOH_2^+$ is small. Therefore in the standard pyridine buffer the C-protonation is faster than O-protonation. However, in a buffer containing a concave pyridine **3**, the protonated nitrogen atoms of the concave pyridines **3** are shielded. This retards the C-protonation and a small concentration of protonated ethanol ($EtOH_2^+$) is sufficient for O-protonation and the Nef-reaction. Thus, the Nef-reaction is possible in a buffered medium, under "softer" conditions than with mineral acids. The reaction was therefore called the *Soft Nef-Reaction*.

A variation of the concentration of concave pyridine **3** and *p*-toluenesulfonic acid (**48**) in the buffers gave further proof [35] that the C-protonation is a *general protonation* [37] while the O-protonation is a *specific protonation* [37] by protonated solvent molecules. The shift from general to specific protonation in the Soft Nef-Reaction is caused by the sterical shielding of the nitrogen atom of the concave pyridines **3**. But also other very large α-substituents like *tert*-butyl groups are sufficient [2,6-di-*tert*-butylpyridine (**49**)] to change the course of the reaction from C- to O-protonation. The soft Nef-Reaction may be carried out with primary and secondary, with cyclic and acyclic nitronate ions (see below).

4.1.2 Substrate Selectivity

The Soft Nef-Reaction was carried out with different nitronate ions **44** and different buffers as shown in Table 2 [38]. As expected the use of *p*-toluenesulfonic acid (**48**) alone resulted in the formation of the Nef-products **47** whereas pyridine (**50**) and 2,6-dimethylpyridine (**51**) buffers gave the nitro compounds **45**.

The Soft Nef-Reaction was possible using the sterically shielded 2,6-di-*tert*-butylpyridine (**49**). But when the concave pyridine **3c** was used, different products (nitro compounds **45** or carbonyl compounds **47**) were found for

Table 2. Competition between C-protonation (formation of nitro compounds **45**) and O-proton-ation (formation of the Nef-product **47**) in the protonation of nitronate ions **44** (s. Scheme 8). The percentage of C-protonated product **45** for different buffers (pyridine/p-toluenesulfonic acid) and unbuffered p-toluenesulfonic acid (**48**) is listed

		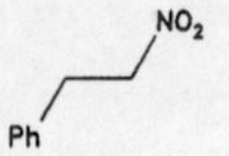 45a	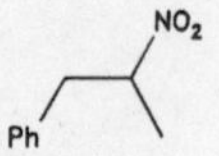 45b	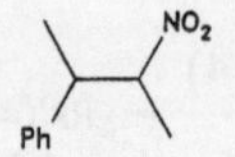 45c	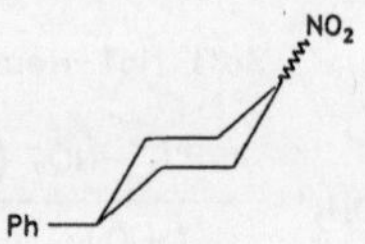 45d
pTsOH	48	< 5	0	0	0
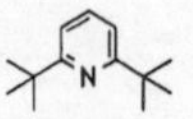 49	49	25	0	0	0
50	50	100	100	100	ca. 75
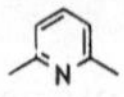 51	51	100	100	90	90 – 95
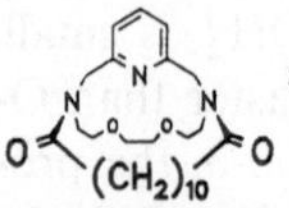 3c	3c	>90	0	15 – 20	25 – 30
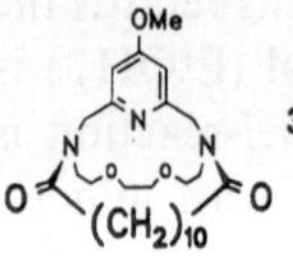 3j	3j	100	100	30 – 70	90

different nitronate ions **44**. The primary nitronate ion **44a** was C-protonated while the secondary nitronate ions **44b–d** yielded the products of the Nef-Reaction (**47**). Concave pyridine **3c** (Table 1) is able to discriminate between primary and secondary nitronate ions **44** in competition reactions: it shows substrate selectivity.

The pH sensitivity of the Nef-reaction (general vs. specific protonation, Section 4.1.1) becomes obvious when the concave pyridine **3j** was used instead of **3c**. Both concave pyridines **3c** and **j** have the same structure; the only difference is the 4-substitution by a methoxy group leading to another basicity. Therefore with the same buffer concentration, the pH of the **3j**-buffer is higher leading to more C-protonation.

4.1.3 Stereoselectivity

The C-protonation of the cyclic anion **44d** gives two stereoisomers: *cis*-**45d** and *trans*-**45d**. The thermodynamics favor the *trans*-compound *trans*-**45d** by 4:1 [39,

40]. But when pyridine buffers were used for the protonation, an increasing degree of 2,6-substitution in the pyridines increased the amount of *cis*-product *cis*-**45d** formed [35, 39]. Because concave pyridines **3** and concave 1,10-phenanthrolines **21** may be regarded as bases with very large α-substituents, concave pyridines **3** and concave 1,10-phenanthrolines **21** were tested for their influence on the stereoselectivity of this reaction [41]. But for the investigated concave bases, hardly any stereoselectivities were found. Maybe the shielding of the protonated pyridine nitrogen atom is too large in these molecules. Fortunately another class of bases was also checked in this reaction: bis-*ortho*-substituted 2-aryl-1,10-phenanthrolines **52**. These molecules are twisted along the aryl-phenanthroline axis and are concave, too, but more shallow.

When these bases **52** were used in buffers for the C-protonation of the cyclic anion **44d**, high stereoselectivities were found as Table 3 shows [42]. Remarkably, the thermodynamically less stable *cis*-product *cis*-**45d** was formed in

Table 3. The protonation of the nitronate ions **44c** and **44d** by protonated 2-aryl-1,10-phenanthrolines **52** leads to *threo/erythro* or *cis/trans* mixtures of the products **45c** or **45d**, respectively (s. Table 2)

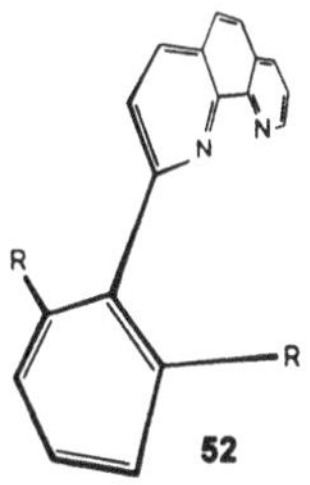

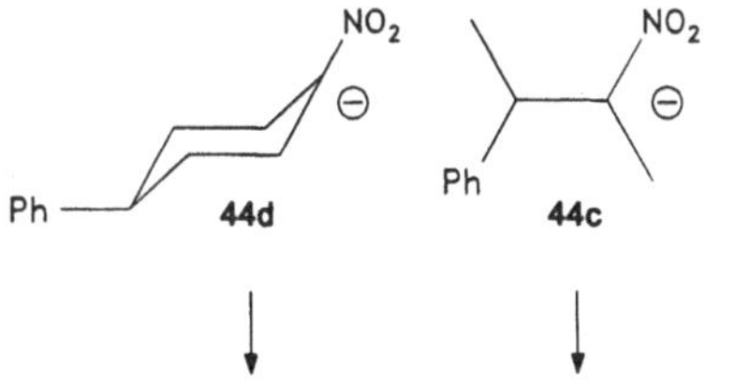

			cis/trans– **45d**	threo/erythro– **45c**
equilibrium			0.2 – 0.25	< 1
pyridines			0.7 – 3.4	0.9 – 1.2
concave pyridines **3**			0.6 – 1.0	–
concave 1,10–phenanthrolines	**21**		0.7	–
52a	R = H		1.5	–
52b	R = Me		8.0	–
52c	R = OMe		8.7	1.0
52d	R = OAc		2.7	10.4
52e	R = OOCTBu		16.1	1.1
52f	R 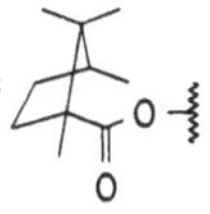 =		13.3	27.4
52g	R 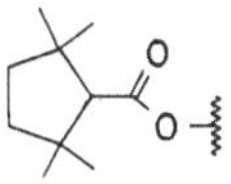 =		12.6	0.9

U. Lüning

excess. In Scheme 10, the two reaction pathways which lead to the *cis*- and *trans*-products *cis*-**45d** and *trans*-**45d** are sketched. In the reaction leading to the *cis*-cyclohexane *cis*-**45d**, the protonated 1,10-phenanthroline **52·H**$^+$ attacks equatorially. Because the small proton is still half bound to the large 1,10-phenanthroline **52** – it is concavely wrapped – the equatorial orientation of this large pseudo-substituent is favored. In Scheme 10, the concavely wrapped proton is drawn as a proton in a circle. The product of this transition state *eq*-**53**, however, is the *cis*-nitro compound *cis*-**45d**, the thermodynamically less stable product. The concave wrapping of the proton is therefore the reason for a contra-thermodynamic protonation. This finding is not restricted to cyclic compounds. An excess of the thermodynamically less stable products was also found for the protonation of the acyclic nitronate ion **44c** [40] (see Table 3).

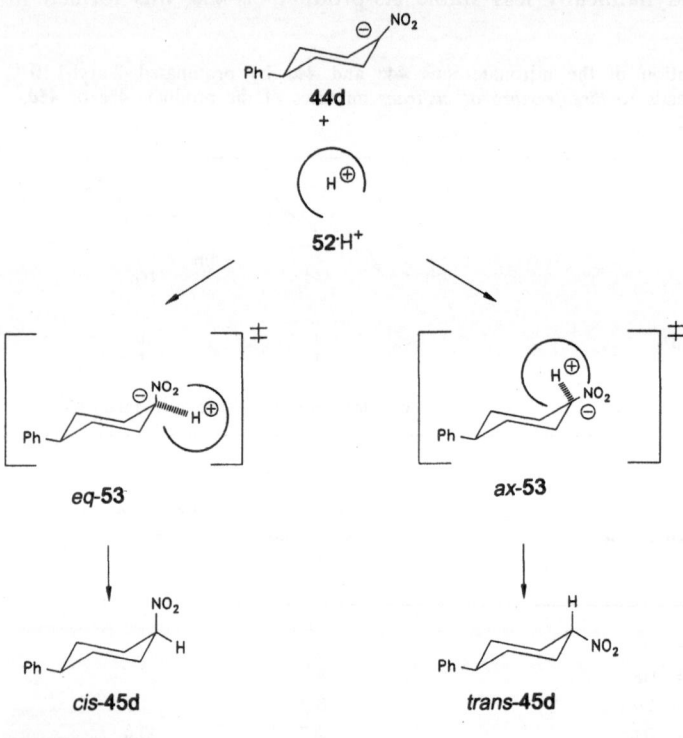

Scheme 10

4.1.4 Protonation of Allyl Anions

The high selectivities found in the protonation experiments of the nitronate ions **44** suggested that also allyl anions **54** can be regioselectively protonated by a general acid protonation. Therefore, some lithium allyl compounds (Structures 6) were generated by deprotonation of alkenes with *n*-butyl lithium.

82

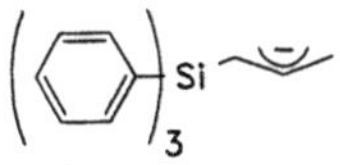

Structures 6

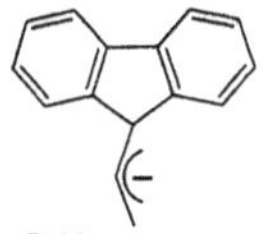

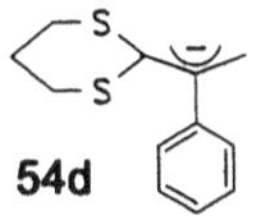

When various acids **55** were used for the protonation of these anions **54**, indeed influences of the nature of the acid **55** on the α,γ-protonation ratios were found [33, 43]. But good selectivities were only found for the triphenylsilylallyl anion (**54a**) (see Table 4). Other anions **54** either had an intrinsic selectivity in extreme favor of one regioisomer, or the α,γ-ratio could not or only slightly be altered by the use of different acids **55**.

Table 4. Regioselectivity of the protonation of the triphenylsilylallyl anion (**54a**) by various acids **55** in diethyl ether [43]

	acid	$\begin{array}{c} Ph_3Si \\ \textbf{54a} \\ \alpha \; : \; \gamma \end{array}$ 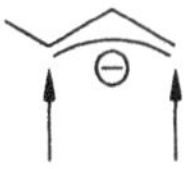
55a	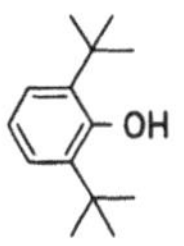 OH	10 : 90
55b	$F_9C_4SO_3H$	35 : 65
55c	H_2SO_4	47 : 53
55d	Ph_3C-H	47 : 53
55e	$(NC)_2CH_2$	48 : 52
55f	tBuOH	49 : 51
55g	PhOH	50 : 50
55h	H_2O	60 : 40
55j	R = Et	84 : 16
55k	R = Me	90 : 10
55l	R = H	90 : 10

Due to the low solubility of the concave pyridines **3** in diethyl ether, the corresponding pyridine buffers could not be compared with the experiments of Table 4. But when the protonations were carried out in other solvents, no influences of the acids (including the acids of Table 4) on the regioselectivity could be found. The exchange of diethyl ether by other solvents caused a color change of the allyl anion solution which indicated different structures for the "anions" in diethyl ether and in other, more polar solvents [44].

Therefore, the attempt to direct the protonation of allyl "anions" systematically by using concave acids is doomed because every change in reagent, reaction conditions and solvent has a strong influence on the aggregates of the anions with the lithium counter ions and these changes cannot be separated from one another.

4.1.5 Summary of Selective Protonation Reactions

Selectivity increases by the use of buffers of concave bases were found for the protonation of nitronate ions **44** (s. Scheme 8). Three interesting results have been found:

- The Soft Nef-Reaction (Nef-reaction in buffers at high pH)
- Substrate selectivity (differentiation between 1° and 2° nitronate ions)
- Contra-thermodynamic stereoselectivity (*cis/trans* or *threo/erythro* selectivity)

In the case of the protonation of allyl anions **54** (Structures 6), no systematic governing of the selectivity is possible at present. Due to the existence of complex mixtures of lithium allyl aggregates, these systems are extremely sensitive to small changes in reaction conditions.

4.2 Model Reactions with Concave Bases

In the previous section, concave bases **3** (Table 1) were used in buffers. Therefore, the reacting species had been the conjugate acids, the protonated concave bases **3·H⁺**. Furthermore, the buffers had at least to be equimolar. Therefore as a second model reaction, a reaction was chosen in which the concave bases **3** operate as a base and in addition as a catalyst. The bases are not used up in the course of the reaction as the protonated conjugate acids were in Sect. 4.1.

There are many reactions in which pyridines are used as bases. However in a large number of reactions only pyridine itself is reactive. α-Substituted pyridines behave differently, e.g. in the catalysis of acylation reactions with acyl chlorides or anhydrides [45]. The sterical hindrance of the α-substituents decelerates reactions in which a pyridine reacts as a nucleophile. A reaction which can be base-catalyzed by α-substituted pyridines is the addition of alcohols to heterocumulenes such as ketenes and isocyanates. Therefore this reaction was investigated as a model reaction for base catalysis by concave pyridines.

4.2.1 Reactivity

In contrast to pyridine catalyzed acylations with acyl chlorides or anhydrides where the pyridine acts as a nucleophile, the base-catalyzed addition of alcohols to heterocumulenes [46] may occur as a general base-catalyzed reaction in which also α-substituted pyridines are active, e.g. 2,6-dimethylpyridine (**51**). Therefore concave pyridines **3** were tested, too. While the addition of ethanol to phenylisocyanate could hardly be accelerated by addition of pyridines [15], the reaction of alcohols with ketenes was accelerated by various pyridines. Two mechanisms (general base catalysis [37, 46c] vs nucleophilic catalysis [37, 46d]) are discussed in the literature and shown in Scheme 11.

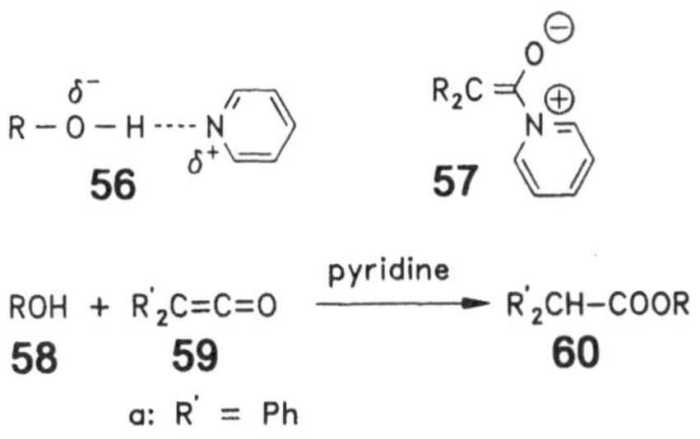

Scheme 11

In a general base catalysis, the pyridine forms a hydrogen bond to an alcohol function (**56**). This causes a polarization and increases the nucleophilicity of the alcohol oxygen thus accelerating the reaction [46c]. The second mechanism postulates a betain intermediate **57** which is formed by a nucleophilic attack of the pyridine on the ketene **59** [46d].

A variety of concave pyridines **3** (Table 1) and open-chain analogues have been tested in the addition of ethanol to diphenylketene (**59a**). Pseudo-first-order rate constants in dichloromethane have been determined photometrically at 25 °C by recording the disappearance of the ketene absorption [47]. In comparison to the uncatalyzed addition of ethanol to the ketene **59a**, accelerations of 3 to 2500 were found under the reaction conditions chosen. Two factors determine the effectiveness of a catalyst: basicity and steric shielding. Using a Brønsted plot, these two influences could be separated from one another. Figure 4 shows a Brønsted plot for some selected concave pyridines **3** and pyridine itself (**50**).

When (concave) pyridines with identical α-substitution but different basicity (caused by 4-substitution) were compared, a linear basicity-reactivity dependence was found. For different types of pyridines the lines run more or less parallel [47]. The slopes of these lines are approx. 0.3. This argues for the association mechanism between the alcohol and the pyridine (via **56**) because for

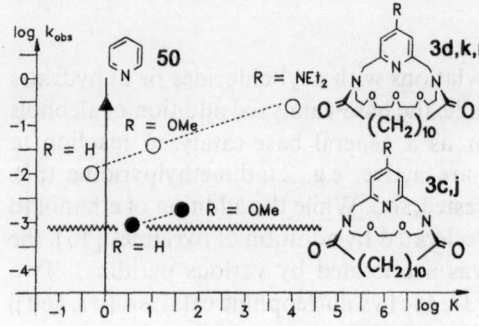

Fig. 4. Brønsted-plot of the logarithms of observed rate constants log k_{obs} for the base catalyzed addition of ethanol to diphenylketene (**59a**) (s. Scheme 11) under standardized conditions vs. a relative basicity log K [31]

mechanisms where a pyridine acts as a nucleophile (via **57**) slopes of approx. 0.8 are expected [48]. By comparing the straight lines at one basicity (e.g. log K = 0), a sterical shielding factor S for the pyridine nitrogen atom for this catalysis could be determined [49].

$$S = \frac{k_{obs}(0)(\text{unhindered pyridine})}{k_{obs}(0)(\text{sterically hind., } \alpha\text{-disubstituted pyridine})}$$

For concave and non-concave pyridines a large variety of S-values were found [47]. α-Unsubstituted pyridines were the most reactive ones but 2,6-dimethylpyridine (lutidine) was only 1.2 times slower. Larger α-substituents led to larger shielding factors: $S = 115$ for 2,6-di-*tert*-butylpyridine (**49**) and 339 for 2,6-diphenylpyridine. The larger concave pyridines **3d**, **3k**, and **3r** (with a longer chain X) catalyzed the reaction rather good ($S = 25$) whereas the smaller bimacrocyclic pyridines **3c** and **3j** (with a shorter chain X) were at least 600 times slower. Comparison with the catalysis by a carboxamide group suggests that the carboxamide bridgeheads are the catalytically active sites in the smaller concave pyridines **3c** and **3j**. In Fig. 4 this amide-catalysis is indicated by a zig-zag line.

In contrast to the pyridine nitrogen atom, the carboxamide functions are *not* located within the concave space of the catalyst **3**. Therefore, only concave pyridines **3** with a distinctly higher reactivity than *N,N*-dimethylacetamide can be considered *concave* catalysts. On the other hand, the basicity dependence of the catalytic activity of the larger concave pyridines **3d**, **3k** and **3r** argue for the pyridine nitrogen atom as the catalytic center. In 4-diethylamino-substituted concave pyridines like **3r**, a further competition is conceivable, the competition between the basicity centers in the concave (pyridine) and convex (diethylamino) position. But the very slow catalysis by *N,N*-diethylaniline argues strongly against a catalysis by the diethylamino group [47]. Concave pyridinebissulfonamides **13** (s. Scheme 3) also catalyzed the addition of alcohols to diphenylketene (**59a**) [47]. With a shielding factor S of 107, their reactivity was between the smaller (**3c**, **3j**) and the larger (**3d**, **3k**, **3r**) pyridine bislactams.

The addition of ethanol to diphenylketene (**59a**) shows that a standard catalyst like pyridine (**50**) may be incorporated into a concave structure and that the resulting concave pyridines **3** are still catalytically active. The reactivity, however, depends on the sterical shielding *and* on the basicity. As shown for the 4-(*N,N*-diethylamino)-substituted larger concave pyridine **3r**, the sterical hindrance by the concave shielding ($S = 25$) can be compensated by introduction of a basicity-increasing substituent like a diethylamino group. With this concave pyridine **3r**, a reactive concave catalyst with a catalytic activity comparable to pyridine itself (**50**) is available. Therefore alcohols **58** which are less reactive than ethanol could also be investigated and the influence of the concave structure on the selectivity of such catalysts in the addition of varying alcohols to ketenes could be studied.

4.2.2 Selectivity

4.2.2.1 1° vs 2° Alcohols

As discussed above, the catalysis of the alcohol addition to diphenylketene (**59a**) (Scheme 11) can be explained by the formation of an alcohol-concave pyridine complex **61** (Structure 7). The differences in catalyses by different concave pyridines **3** (Table 1) may be explained with the varying ring sizes in the catalysts (see above: sterical shielding S). Ethanol may form more easily a hydrogen bond to the pyridine nitrogen atom of a larger concave pyridine **3** and the polarized oxygen atom in the resulting complex will also be more accessible for diphenylketene (**59a**). But when the variation of the size of the concave pyridine **3** has an influence on the rate of catalysis, one can also expect rate differences when the catalyst is kept constant but the alcohol **58** is varied, i.e. when 1°, 2° or 3° alcohols are compared. Rate constants for standardized reactions of diphenylketene (**59a**) with ethanol and isopropanol were measured [15, 47]. The quotients of the rate constants k_{obs} [EtOH]/k_{obs}[i-PrOH] are listed as calculated selectivities 1°/2°$_{calc}$ in Table 5.

There is an inherent selectivity for the addition of a 1° and a 2° alcohol to diphenylketene (**59a**): the selectivity of the uncatalyzed reaction is not unity. Also the bridgeheads of the concave pyridines **3** (Table 1) and **13** (s. Scheme 3)

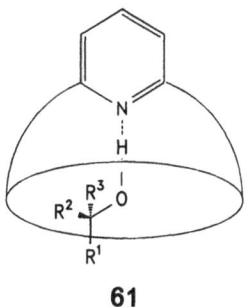

61

Structure 7

Table 5. Calculated selectivities $1°/2°_{calc}$ for the addition of ethanol and isopropanol to diphenylketene (**59a**) (s. Scheme 11) (quotients of the rate constants k_{obs} [EtOH]/k_{obs} [iPrOH]) [15, 47]

	$1°/2°_{calc}$
uncatalyzed	1.5
Me$_2$N–COMe **62**	1.9
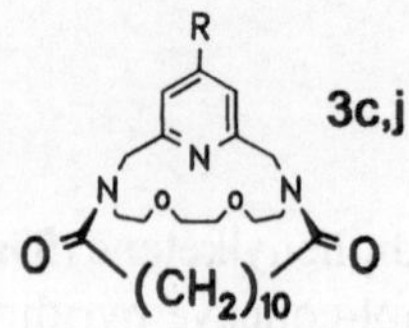 **3c,j**	1.8 – 1.9
Et$_2$N–SO$_2$Me **63**	1.5
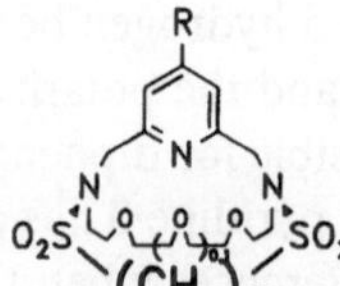 **13**	1.8 – 1.9
 **50**	3.8
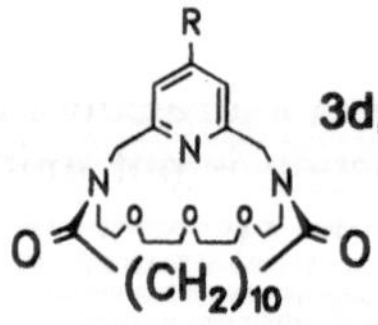 **3d,k,r**	5.3 – 5.9

may catalyze the addition [47], the selectivities of carboxamide (**62**) and sulfonamide (**63**) (Table 5) catalyses are also larger than unity. The known fact (see above, [47]) that small concave pyridines **3c** and **3j** are catalyzing via their carboxamide functions is supported by their selectivities. Also in the case of bimacrocyclic pyridine sulfonamides **13**, the selectivities did not vary much from those obtained for the sulfonamide **63** alone. But when a set of larger concave pyridines **3d**, **3k** and **3r** was used, good selectivities $1°/2°_{calc}$ were found. The incorporation of the pyridine into the bimacrocyclic structure in fact influenced the catalyses of the isopropanol addition more than it did for ethanol. The similar selectivities $1°/2°_{calc}$ for **3d**, **3k** and **3r** which differ strongly in their basicity also show that the shape of the concave pyridine **3** is responsible for this selectivity and that the basicity has hardly any effect.

These results are confirmed by direct competition experiments where diphenylketene (**59a**) was added to a solution of the catalyst which contained a mixture of alcohols [50]. While the direct competition experiments gave the same trend (**3d**, **3k**, **3r** are the most selective catalysts) the absolute values of $1°/2°_{comp}$ were much larger than the $1°/2°_{calc}$ selectivities:

Pyridine (**50**)	$1°/2°_{calc} = 3.8$	$1°/2°_{comp} = 4.2$
3c, **3j**:	$1°/2°_{calc} = 1.8 - 1.9$	$1°/2°_{comp} = 3.5 - 4.6$
3d, **3k**, **3r**:	$1°/2°_{calc} = 5.3 - 5.9$	$1°/2°_{comp} = 8.3 - 11.5$

The differences between $1°/2°_{calc}$ and $1°/2°_{comp}$ are probably caused by the changes in the reaction media. In unpolar media, alcohols form telomeric, hydrogen bonded complexes (rings or chains) whose nature depend on the concentration(s) and the type of alcohol(s) [51]. When a competition experiment is compared to the addition of a single alcohol either the *overall* alcohol concentration can be chosen identical, or *each* alcohol concentration may be the same leading to a higher overall concentration. Indeed, in each case different selectivities were found [15].

The data in Table 5 show that concave pyridines **3** are able to differentiate alcohols. In synthesis, this capability is important when OH-groups within *one* molecule shall be differentiated. Therefore, intramolecular selectivities were also determined. Table 6 lists the results of intramolecular competition reactions for 1,2-propanediol (**64**) $1°/2°_{intra}$. Again, the small concave pyridine **3c** only showed a small selectivity whereas the large concave pyridines **3d** and **3r** were the most selective catalysts. With a selectivity of 15 (**3r**), a selectivity range is reached which may be useful for applications because more than 93% of the functionalized OH-groups are primary ones.

Table 6. Intramolecular competition constants $1°/2°_{intra}$ for the base catalyzed addition of diphenylketene (**59a**) (s. Scheme 11) to 1,2-propanediol (**64**) to give mono-acylated propanediols **65**

	3c	49	50	3d	3r
64 R¹, R² = H 65 R¹, R² = CO–CHPh₂, H or H, CO–CHPh₂				R = H	R = NEt₂
$1°/2°_{intra}$	3	$5 - 7$ [a]	7	$8 - 12$ [a]	15

[a] Depending on the concentration of the diol **64**.

89

4.2.2.2 2° vs 2° Alcohols

Of even greater importance than a 1°/2° selectivity is the differentiation between secondary alcohols, e.g. in sugars and other natural products (chiral pool).

In a first set of experiments [52], the methyl ester of cholic acid **66a** (Structures 8) was reacted with diphenylketene (**59a**) in the presence and in the absence of catalysts [pyridine (**50**), concave pyridine **3r**]. But in all experiments, only the equatorial group in 3-position could be acylated (**66b**). No functionalization of the axial OH-groups in 7- and 12-position was found (**66c, 66d**).

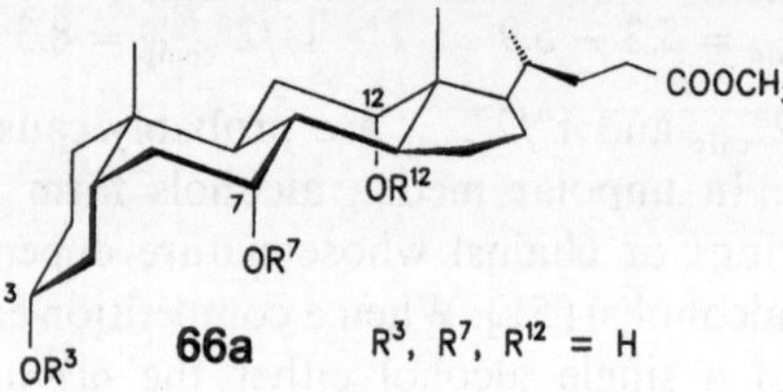

66a $R^3, R^7, R^{12} = H$

66b	**66c**	**66d**
$R^3 = COCHPh_2$	$R^7 = COCHPh_2$	$R^{12} = COCHPh_2$
$R^7, R^{12} = H$	$R^3, R^{12} = H$	$R^3, R^7 = H$

Structures 8

Table 7. Yields of acylated glucose derivatives **67b–d** in the base catalyzed reaction of the glucose derivative **67a** with diphenylketene (**59a**) (s. Scheme 11). The only other glucose derivative **67** found was unreacted educt **67a**

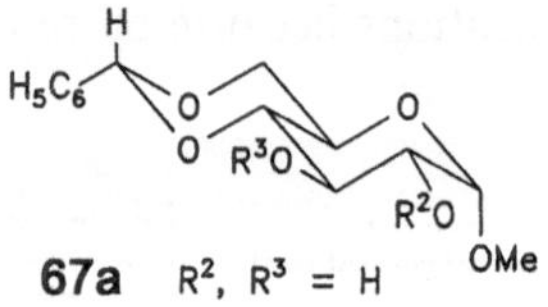

67a $R^2, R^3 = H$ OMe

	67b $R^2 = H$ $R^3 = Ph_2CH–CO$	**67c** $R^2 = Ph_2CH–CO$ $R^3 = H$	**67d** $R^2 = Ph_2CH–CO$ $R^3 = Ph_2CH–CO$
uncatalyzed	36	42	3
pyridine **50**	12.5	50	12.5
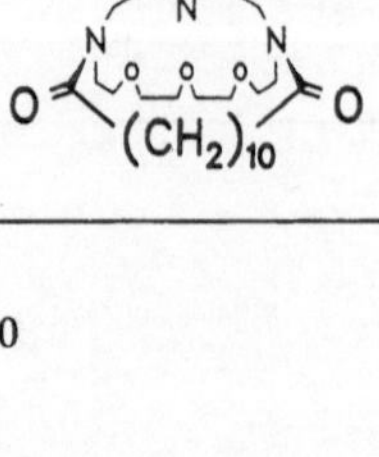 **3r**	7	67	n.d.

Therefore in a second set of experiments, the reaction between diphenylketene (**59a**) and a glucose derivative **67a** with two equatorial OH-groups was investigated. Table 7 lists the product distribution for the reaction of one equivalent of **67a** with one equivalent of diphenylketene (**59a**) in the presence and absence of catalysts.

As expected, axial OH groups were easier to differentiate from equatorial ones than equatorial OH groups from one another. In the case of methyl cholate **66a**, a standard reagent (pyridine, **50**) does a good job. But in the glucose derivative **67a**, the two equatorial OH groups are much more similar to one another. Therefore it is not surprising that they react with almost the same rate in the uncatalyzed reaction. When pyridine (**50**) was used as catalyst, the acylation of the 2-position (**67c**) was preferred by a factor of 4 but also a bis-acylated product **67d** was formed. Concave pyridine **3r** showed the best results. With a selectivity of 9 : 1, the 2-acylated product **67c** was formed and no diacylated product **67d** could be determined.

The selectivity studies show that concave pyridines **3** (Table 1) not only catalyze the addition of alcohols to ketenes, but they may also differentiate between different OH groups in inter- and intramolecular competitions. They are not only *reactive* but also *selective*. First experiments with a chiral concave 1,10-phenanthroline show that enantioselectivity is also possible [20]. Structure 9 shows the concave 1,10-phenanthroline **21g** which catalyzes the addition of *R*-1-phenylethanol (*R*-**68**) to diphenylketene (**59a**) 20% faster than the addition of the S-enantiomer *S*-**68**.

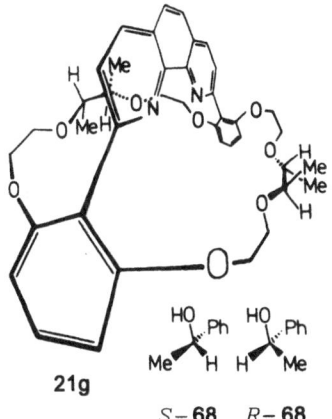

21g

S – **68**   *R* – **68**

Structure 9

4.3 General Conclusions from the Model Reactions

The model reactions of Sects. 4.1 and 4.2. have shown that the reactivity of a pyridine derivative may also be found for their concave analogues. The concave pyridines 3 could be used as acids or as bases and improved selectivities were found due to the concave shielding of the pyridine.

When the concave reagents are compared to other reactions in Supramolecular Chemistry a distinct difference must be noted: Most other approaches try to bind the substrate in a host first. Then this complex reacts with a reagent which either is present in solution or attached to the host. For concave reagents and concave catalysts, however, there is no need for binding of the educt. In contrast, the protonation reactions can be interpreted as a reagent (H^+) host complex.

5 Recycling

As outlined in Sect. 2, various classes of concave acids and bases may be synthesized in gram quantities. But the syntheses are multistep sequences and the yields are often limited by the macrocyclization steps. Therefore, for a practical use, these reagents are quite "expensive" and recovery and recycling is necessary.

In order to improve such a work-up, two concave pyridines 3 have been attached to a polymeric back-bone in order to recover the concave base by simple filtration. A spacer has been attached to the convex outside of the pyridine 3, i.e. in 4-position of the pyridine ring. As spacer, an ethyleneglycol unit was chosen because the spacer should not be too long to avoid a folding of the hollow cave of the concave pyridine onto the polymeric backbone. On the other hand, a spacer reduces interactions between the backbone and the concave pyridine and enhances partial solvation.

The spacer-carrying, benzyl-protected concave pyridines 3n and 3o were synthesized as described in Sect. 2.1.1 [16, 17]. Deprotection to give 3l and 3m (Scheme 12) was possible by hydrogenation. Sodium hydride proved to be the reagent of choice with which the glycol OH group could be deprotonated for alkylation without damaging the bimacrocyclic structure. The concave pyridines 3l and 3m with different ring sizes have been attached to a Merrifield polymer 69. FT-IR spectroscopy, elemental analyses and titration characterized the functionalized polymers 70 (Scheme 12). Quantitative analyses showed that the modified polymers 70 contained 10% (w/w) of concave pyridine 3n or 3o, respectively [16, 17].

The modified polymer 70a with the larger concave pyridine was then tested in the base catalyzed addition of ethanol to diphenylketene (59a) and proved to be catalytically active [17]. To compare the polymer bound concave pyridine 70a to the corresponding free concave pyridine 3k (MeO-substituted in 4-position of the pyridine ring), the same quantities of pyridine units were used, in solution (3k) or in suspension (70a) respectively. The polymer bound catalyst 70a catalyzed 2–3 times slower than the analogous 4-methoxy-substituted

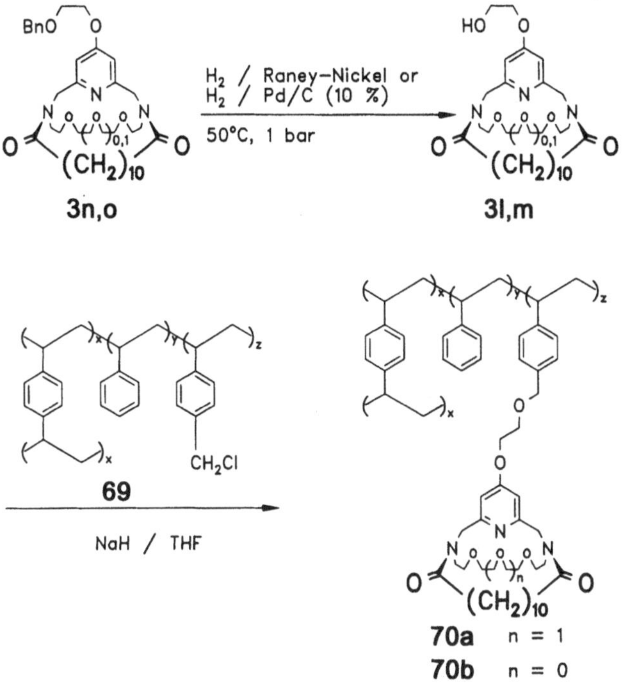

Scheme 12

concave pyridine **3k**. Whether this result is caused by the differences between a heterogeneous and a homogeneous catalysis, or whether the polymer bound concave pyridine **70a** is not as reactive as soluble ones, cannot be answered at present.

6 Outlook

6.1 Concave Acids and Bases

Besides concave pyridines, concave 1,10-phenanthrolines and concave benzoic acids, a large number of other concave acids and bases are conceivable containing other acidic or basic groups in the concave position. The central question for such new concave reagents is: how can the new functionality be incorporated into the bimacrocyclic structure, and how can a concave orientation be assured?

Many supramolecular systems contain endo-functionalities. But in most cases, the concave positioning of the functional group is not assured [53].

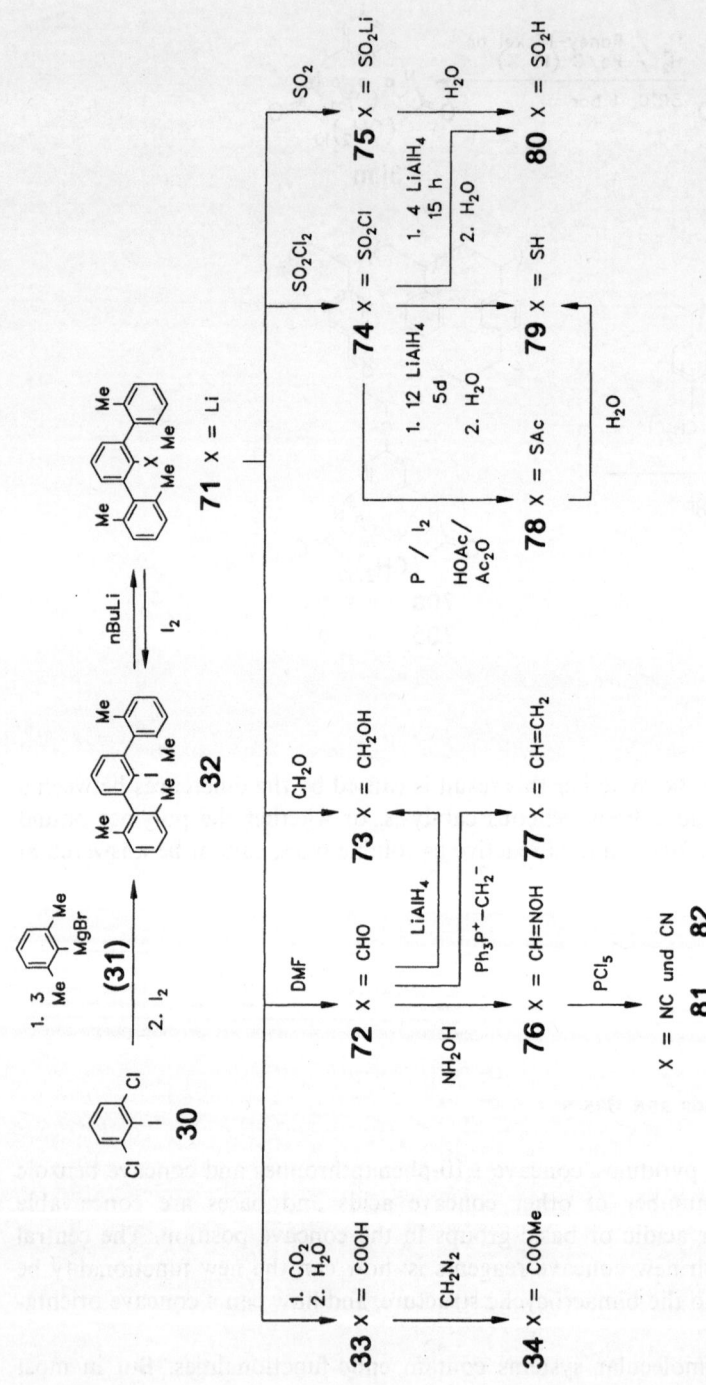

Scheme 13

Therefore structural units as the *m*-terphenyl moiety, which was used in the concave benzoic acids **38**, would be useful if new functionalities could be placed in the 2'-position. In the case of the benzoic acid **33**, the functional group (COOH) was incorporated by the reaction of the 2'-lithio compound **71** with CO_2. But other electrophiles may also be used here [28]. Scheme 13 lists new *m*-terphenyls **72–75** which were obtained when electrophiles other than CO_2 were used. Standard group transformations transferred these products into other 2'-substituted *m*-terphenyls **76–82** which are also listed in Scheme 13.

In the thiol **79** and the sulfinic acid **80**, two new acidic functionalities are now present in the 2'-position of a *m*-terphenyl system and should be incorporable into concave structures by a double bridging of the *m*-terphenyl. For the acetyl-protected thiol **78**, the corresponding concave thiol acetate **42** (Scheme 7) could be obtained after tetrabromination of the four methyl groups and bridging with *m*-phenylenedithiol (see Sect. 2.3).

6.2 Other Concave Reagents

When acids and bases can be altered in their reactivity and selectivity by incorporating them into concave structures, this should also be possible for other functionalities. In addition, the functional groups used in the known concave acids and bases do not have only acid/base properties. This allows them to be used not only as acids or bases but also for other purposes.

6.2.1 Hydrogen Atom Transfer

While in concave acids and bases, protons (hydrogen cations) are transferred, a thiol can also transfer hydrogen atoms. Therefore *m*-terphenyl-2'-thiol derivatives **84** (Structures 10) should be investigated for their use in hydrogen atom transfer reactions and compared with thiophenol. Thus for instance, the regioselectivity of hydrogen atom transfer to allyl radicals **83** may be governed by the concave shielding of a concave thiol **84** [54].

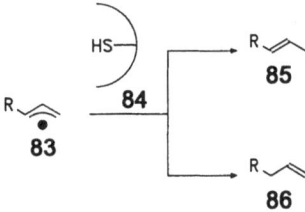

Structures 10

As a precursor, a concave thiol acetate **42** (Scheme 7) has already been synthesized (see Sect. 2.3).

6.2.2 Redox Reagents

The free electron pair(s) in the concave pyridines **3** (Table 1), **13** (s. Scheme 3) and **29** (s. Scheme 5) and especially in the concave 1,10-phenanthrolines **11** (s. Scheme 2) and **21** (Structures 3) are not only able to bind a proton, they may also be used to coordinate a metal ion. For concave 1,10-phenanthrolines **11** and **21**, transition metal complexes **87** (Structure 11) have already been generated [18, 55]. They form readily in acetonitrile solution with binding constants of 10^4–10^7 and larger. Of great importance is the nature of the chains X in the concave 1,10-phenanthrolines **21** (Structures 3). Pure aliphatic chains lead to smaller association constants than polyether chains.

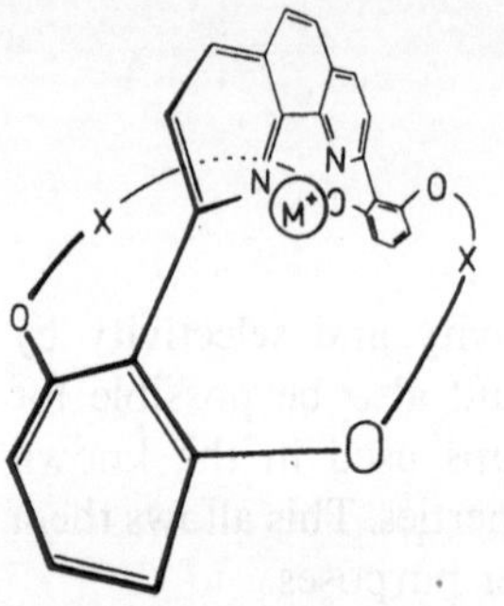

87

Structure 11

If the transition metal could exist in two different oxidation states in the complex **87**, one would have a concave redox reagent which could be useful for instance in epoxidation reactions [56]. The concave shielding of the metal ion should influence the regio-, stereo- and, if the concave 1,10-phenanthroline **21** is chiral, enantioselectivity of an epoxidation.

Acknowledgements. This work was sponsored by the *Fonds der Chemischen Industrie*, the *Deutsche Forschungsgemeinschaft*, the *Wissenschaftliche Gesellschaft Freiburg* and Prof. Dr. *C. Rüchardt*. I am very thankful to my co-workers whose names appear in the references.

7 References and Footnotes

1. Blum Z, Lidin S, Anderson S (1989) Angew Chem 100: 995; Int Ed Engl 27: 953
2. Some books and reviews on Supramolecular Chemistry and Host Guest Chemistry: (a) Weber E,

Vögtle F (eds) (1992) Macrocycles. Top Curr Chem 161; (b) Weber E (ed) (1993) Supramolecular Chemistry I-Directed synthesis and molecular recognition. Top Curr Chem 165. (c) Vögtle F (1989) Supramolekulare Chemie. Teubner, Stuttgart; (1991) Supramolecular Chemistry. Wiley, Chichester. (d) Vögtle F, Weber E (eds) (1985) Host guest complex chemistry-Macrocycles, synthesis, structures, applications. Springer, Berlin Heidelberg New York (e) Schneider HJ, Dürr H (eds) (1991) Frontiers in supramolecular chemistry and photochemistry. VCH, Weinheim. (f) Atwood JL (Ed) (1990) Inclusion phenomena and molecular recognition. Plenum, New York; (g) Atwood JL, Davies JED, MacNicol DD (eds) (1991) Key organic host systems (vol 4 in Series Inclusion Compounds). Oxford University Press, Oxford; (h) Dietrich B, Viout P, Lehn JM (1992) Macrocyclic chemistry. VCH, Weinheim; (i) Izatt RM, Christensen JJ (eds) (1979, 1982, 1986) Progress in macrocyclic chemistry vol 1–3, Wiley, New York; (j) Patai S, Rappoport Z (eds) (1989) Crown ethers and analogs (Updates from the Chemistry of Functional Groups), Wiley, Chichester; (k) Inoue Y, Gokel GW (eds) (1990) Cation binding by macrocycles. Marcel Dekker, New York; (l) Lindoy LF (1989) The chemistry of macrocyclic ligand complexes. Cambridge University Press, Cambridge; (m) Cooper SR (ed) (1992) Crown compounds: Toward future application. VCH, Weinheim; (n) Inoue Y, Gokel GW (eds) (1990) Cation binding by macrocycles. Marcel Dekker, New York. (o) Vicens J, Böhmer V (eds) (1991) Calixarenes: a versatile class of macrocyclic compounds (Topics in Inclusion Science, vol 3). Kluwer, Dordrecht; (p) Gutsche CD (1989) Calixarenes (Stoddart JF (ed) Monographs in Supramolecular Chemistry). Royal Society of Chemistry, Cambridge; (q) Lectures from the 17th International Symposium on Macrocyclic Chemistry (1993) Pure Appl Chem 65: Issue 3, p 355–594. (r) Cram DJ, Cram JM (1994) Container Molecules and Their Guests (Stoddart JF (ed) Monographs in Supramolecular Chemistry). Royal Society of Chemistry, Cambridge; (s) For an excellent summary of other leading Refs see: Schneider HJ (1991) Angew Chem 103: 1419; Int Ed Engl 30: 1417

3. (a) Keehn PM, Rosenfeld SM (eds) (1983) Cyclophanes (Organic Chemistry, vol 45). Academic Press, New York; (b) Vögtle F (1990) Cyclophan-Chemie. Teubner, Stuttgart (c) Diederich F (1991) Cyclophanes (Stoddart, JF (ed) Monographs in Supramolecular Chemistry). Royal Society of Chemistry, Cambridge

4. (a) Binding of cations and anions: Izatt RM, Pawlak K, Bradshaw JS, Bruening RL (1991) Chem Rev 91: 1721–2085; (b) Binding of neutral molecules: Izatt RM, Bradshaw JS, Pawlak K, Bruening RL, Tarbet BJ (1992) Chem Rev 92: 1261; (c) Energetics of binding: [2r]

5. Some examples for host-guest systems with a stabilization of the transition state: (a) Stauffer DA, Barrans Jr RE, Dougherty DA (1990) Angew Chem 102: 953, Int Ed Engl 29: 915; (b) Schmidtchen F (1986) Top Curr Chem 132: 117

6. An example where two different substrates are bound to a host to react with one another: Kelly TR, Zhao C, Bridger GJ (1989) J Am Chem Soc 111: 3744

7. For an early example of functional groups in concave position, see: Helmchen G, Schmierer R (1981) Angew Chem 93: 208; Int Ed Engl 20: 205. Here a dienophil is covalently bound in a concave position. The selectivity of its Diels-Alder reaction is determined by the concave environment

8. Concave acids and bases also exist as conjugate bases and acids. Therefore each concave acid and base can be used as acid *and* as base

9. The term *bimacrocycle* will be used instead of macrobicycle because bimacrocycles often also contain smaller rings (e.g. aryl rings). Then they are macro*polycyclic* although only two *macrocycles* exist. (In addition, no bicycle company has offered a sponsorship yet!)

10. For the description of *in-in, in-out-* or *out-out-*isomers see: Franke J, Vögtle F (1985) Angew Chem 97: 224; Int Ed Engl 24: 219

11. Lüning U (1987) Liebigs Ann Chem: 949

12. (a) Lüning U, Baumstark R, Müller M (1991) Liebigs Ann Chem: 987; (b) Lüning U, Baumstark R, Peters K, v Schnering HG (1990) Liebigs Ann Chem: 129

13. Dietrich B, Lehn JM, Sauvage JP, Blanzat J (1973) Tetrahedron 29: 1629

14. (a) Fenton DE, Vigato PA (1988) Chem Soc Rev 17: 69; Fenton DE (1986) Pure Appl Chem 58: 1437, and references cited therein; (b) Gerbeleu NW, Arion WB (1990) Templatniji Sintez Makrotsiklitsheskich Soedineniji (Russ), Academy of the Moldavian SSR

15. Schyja W, Thesis, Universität Freiburg, Germany (in preparation)

16. Lüning U, Gerst M (1992) J Prakt Chem/Chem-Ztg 334: 656

17. Hacker W (1993) Thesis, Universität Freiburg, Germany

18. Lüning U, Müller M (1989) Liebigs Ann Chem: 367

19. Rotational barrier for R-SO$_2$-NMe$_2$: 21–25 kJ/mol: Remizov AB, Butenko GG (1979) Zh Strukt Khim 20: 63; (1979) Chem Abstr 91: 4893v
    Rotational barrier for R-CO-NMe$_2$: 80–92 kJ/mol; Gutowsky HS, Holm CH (1956) J Chem Phys 25: 1228; Duffy EM, Severance DL, Jorgensen WL (1992) J Am Chem Soc 114: 7535; See also [11, 12b]
20. (a) Müller M (1991) Thesis, Universität Freiburg, Germany; (b) Lüning U, Müller M, Gelbert M, Peters K, von Schnering HG, Keller M (1994) Chem Ber 127: in press
21. (a) Dietrich-Buchecker CO, Sauvage JP (1987) Chem Rev 87: 795; (b) Dietrich-Buchecker CO, Sauvage JP (1983) Tetrahedron Lett 24: 5091
22. Lüning U, Müller M (1990) Chem Ber 123: 643
23. Spitzner D (1992) in Houben-Weyl Methoden der Organischen Chemie 4th Ed, E7b: Hetarene II (Ed: Kreher R), p 588ff, Thieme, Stuttgart
24. When compared to phenyl lithium, the addition of 2,6-dimethoxyphenyl lithium to 1,10-phenanthroline is more difficult [22]. But even the introduction of non-hindered lithium organic compounds to 2-phenylpyridine gives only small yields: (a) Gilman H, Edwards JT (1953) Can J Chem 31: 457; (b) Overberger CG, Lombardino JG, Hiskey R (1957) J Am Chem Soc 79: 6430
25. Lüning U, Baumstark R, Schyja W (1993) Tetrahedron Lett 34: 5063
26. (a) Grewal RS, Hart H, Vinod TK (1992) J Org Chem 57: 2721, and Refs Cited therein; (b) Vinod TK, Hart H (1988) J Am Chem Soc 110: 6574; (c) Vinod T, Hart H (1990) J Org Chem 55: 881
27. (a) Lüning U, Wangnick C, Peters K, v Schnering HG (1991) Chem Ber 124: 397; (b) Lüning U, Wangnick C (1992) Liebigs Ann Chem: 481
28. Lüning U, Baumgartner H (1993) Synlett, 571
29. (a) Fastrez J (1987) Tetrahedron Lett: 419; (b) Fastrez J (1989) J Phys Chem 93: 2635
30. Successful mass spectrometry of concave acids and bases was possible in most cases by EI, 70 eV. But in some cases, only other MS techniques gave molecular peaks: field desorption [12a]; FAB-MS (m-nitrobenzyl alcohol) [27b]; EI, 12 eV [28]
31. The resulting log K scale runs parallel to the pK_a scale with log K (thymol blue) = 0. For a definition see [18]
32. Baumstark R (1991) Thesis, Universität Freiburg, Germany
33. Wangnick C (1991) Thesis, Universität Freiburg, Germany
34. QCPE program No. 429 by Connolly ML, used with Chem-X, developed and distributed by Chemical Design Ltd, Oxford, England
35. Lüning U, Baumstark R, Müller M, Wangnick C, Schillinger F (1990) Chem Ber 123: 221
36. Review: Pinnick HW (1990) Organic Reactions 38: 655
37. The term general and specific protonation is used in the same sense as the terms general and specific acid catalyses are used: specific means a specific protonated solvent molecule is the reacting species while general means that in general all acids in solution contribute to the reaction
38. Lüning U, Schillinger F (1990) Chem Ber 123: 2073
39. Zimmerman HE, Mariano PS (1968) J Am Chem Soc 90: 6091
40. Lüning U, Müller M (1992) Angew Chem 104: 99; Int Ed Engl 31: 80
41. The pyridine/pyridinium buffers were mixed in such a manner that the pH was high enough to avoid the Soft Nef-Reaction
42. In the protonation of nitronate ion **44d**, a solvent dependence of the stereoselectivity was found. Alcohols gave the highest selectivities [20, 40]
43. (a) Kümmerlin M (1993) Thesis, Universität Freiburg, Germany; (b) Lüning U, Wangnick C, Kümmerlin M (1994) Chem Ber 127: in press
44. Hünig S, Reichelt H (1986) Chem Ber 119: 1772
45. The hydrolysis of acetic anhydride is not catalyzed by 2,6-dimethylpyridine: (a) Butler AR, Gold V (1961) J Chem Soc: 4362; (b) Deady LW, Finlayson WL (1983) Aust J Chem 36: 1951
46. (a) Franck A (1985) Thesis, Universität Freiburg, Germany; (b) Blake P (1980) in The Chemistry of Ketenes, Allenes and Related Compounds. Part 1 (Ed Patai S), Chapter 9, p 309, Wiley, Chichester; (c) Tille A, Pracejus H (1967) Chem Ber 100: 196; (d) Jähme J, Rüchardt C (1981) Angew Chem 93: 919; Int Ed Engl 20: 885; (e) Pracejus H, Leška J (1966) Z Naturforsch 21B: 30
47. Lüning U, Baumstark R, Schyja W (1991) Liebigs Ann Chem: 999
48. Johnson SL (1967) Adv Phys Org Chem 5: 237. Slopes of ≤ 0.5 argue against nucleophilic reactions, see also [46a]
49. In some cases, only one pyridine of its class was investigated. In these cases, a slope of 0.3 was assumed for the comparison with other pyridines

50. For $1°/2°_{calc}$ and $1°/2°_{comp}$, the concentrations of each alcohol were chosen 50 mM
51. Telomers of alcohols in unpolar media have been investigated: (a) Symons MCR (1983) Chem Soc Rev 12: 1; (b) Landeck H, Wolff H, Götz R (1977) J Phys Chem 81: 718; (c) Bartczak WM (1979) Ber Bunsenges Phys Chem 83: 987; (d) Lammiman SA, Satchell RS (1972) J Chem Soc Perkin Trans II 2300. In mixtures of alcohols, even more complex equilibria are expected
52. Lüning U, Schyja W, Gürtesch L (1993) unpublished results
53. (a) Whitlock BJ, Whitlock HW (1990) J Am Chem Soc 112: 3910. Like in **3**, bimacrocyclic pyridines were synthesized but the spacer between the pyridine ring and aryl bridgeheads were three atoms long allowing an considerably higher flexibility than in **3**.
    (b) Rodriguez-Ubis JC, Alpha B, Plancherel D, Lehn JM (1984) Helv Chem Acta 67: 2264. In addition to the flexibility of a bipyridyl system, in all compounds described here basic N-bridgehead atoms were used
    (c) Cram DJ, Weiss J, Helgeson RC, Knobler CB, Dorigo A, Houk KN (1988) J Chem Soc Chem Commun: 407. Polymacrocyclic 1,10-phenanthrolines based on cyclotriveratrylene derivatives are described. But a concave orientation of the 1,10-phenanthroline-N-atoms is not assured
    (d) Böhmer V, Vogt W (1993) Pure Appl Chem 65: 406. A calixarene is bridged by a bis-*ortho*-methylenephenol. But the phenol OH is not located within the cavity
    A variety of molecules with a functional group located in a cleft have been synthesized: (a) Zimmerman SC, Zeng Z, Wu W, Reichert DE (1991) J Am Chem Soc 113: 183 ("Molecular tweezers"); (b): Rebek Jr J, (1990) Angew Chem 102: 261; Int Ed Engl 29: 245 (Clefts based on derivatives of "Kemp's acid"). In both classes, a functional group is located in a slot. But the shielding is only from the top and from the bottom, not from the sides.
54. Also an increase in the stereoselectivity of hydrogen transfer is conceivable. For stereoselective hydrogen atom transfer to 2,3-disubstituted succinic anhydrid-2-yl radicals see: (a) Giese B, Meixner J (1977) Tetrahedron Lett: 2783; (b) Giese B, Kretzschmar G (1984) Chem Ber 117: 3175; (c) Giese B (1989) Angew Chem 101: 993; Int Ed Engl 28: 969. Stereoselective formation of *cis*-alkenes via H·-transfer to 2-substituted vinyl radicals; see (c); (d) Lachhein S (1982) Thesis, Technische Hochschule Darmstadt, Germany
55. Gelbert M, Thesis, Universität Freiburg, Germany, in preparation
56. By substitution of porphyrins, concave metal complexes have already been synthesized which show increased selectivities in epoxidation reactions, e.g. (a) Suslick KS, Cook BR (1987) J Chem Soc Chem Commun: 200; (b) Breslow R, Brown AB, McCullough RD, White PW (1989) J Am Chem Soc 111: 4517

Molecular Recognition of Organic Acids and Anions – Receptor Models for Carboxylates, Amino Acids, and Nucleotides

Christian Seel, Amalia Galán, and Javier de Mendoza

Departamento de Química Orgánica C-I, Universidad Autónoma de Madrid, E-28049 Madrid, Spain

Table of Contents

The roots of synthetic molecular recognition and supramolecular chemistry itself lie in the investigation of organic ligands for metal cations, and the amount of research done in this field is overwhelming. For a long time, however, far less interest has been focussed on the complexation of anions even if the primal strategies were comparable, and only in recent years have the advances made been increasingly dynamic. In particular receptor molecules for biorelevant species such as amino acids and nucleotides are primary research targets, not least for their potential applications in medicine. In this article the developments which have been made here so far will be summarized and the attentive reader might notice that most of the primary literature cited is less than five years old. The emphasis has not been laid upon structural characteristics or synthetic strategies but rather on the effects the new host molecules give rise to and the functions they may have.

Topics in Current Chemistry, Vol. 175
© Springer-Verlag Berlin Heidelberg 1995

1 Introduction

Anions display important roles in biology. Amino acids, peptides, and nucleotides are representative examples of organic anions in living organisms. Several inorganic anions, such as nitrate, carbonate, sulfate, and chloride, are also present in large amounts in biological systems. Despite these qualities, the coordination chemistry of anions has only recently received attention, in sharp contrast to the far more advanced development of the corresponding coordination chemistry of cations. Among the reasons behind this unequal treatment are the following:

1. Anions are bulkier than cations and have larger ionic radii than cations of the same atomic (molecular) weight and number of charges. This is why they are also more polarizable.

2. Anions are more strongly solvated than cations of comparable size. For example, the solvation energy for the potassium cation (ionic radius 133 pm) is 80.6 kcal/mol, whereas it is 103.8 kcal/mol for the fluoride anion (ionic radius 136 pm). This means that a severe energy penalty has to be paid to desolvate anions prior to complexation, and therefore the stability constants of anion complexes are comparatively smaller. Hydrogen bonds with a large number of water molecules are the main reason for this increased solvation energy. Thus, hydrogen bonding constitutes a very important aspect of the coordination chemistry of anions.

3. The miscellaneous geometries of anions, ranging from spherical (Cl^-) to linear (N_3^-), planar (NO_3^-, CO_3^{2-}), tetrahedral (PO_4^{3-}), or octahedral (PF_6^-), contrasts with the usual spherical shape of most cations.

4. Most anions are protonable species, whose existence, within a given pH range, is governed by the laws of acid-base equilibria.

Artificial receptors for anions should contain cationic or electron-deficient binding sites to complement and neutralize the negative charges of these substrates. Cationic centers can be included in the covalent framework of the receptor (ammonium, guanidinium, phosphonium, etc.) and may also incorporate hydrogen bond donor groups to increase the binding strength of the complex or to achieve a better selectivity. A second possibility is to make use of the cation affinity of the parent ligands similar to the method of substrate binding in metallo enzymes. In these so-called "cascade complexes" the cationic substructures serve as anchor groups for the anionic guests, fixing it through salt bridges in a multicomponent complex. Finally, since most anions contain lone pairs not involved in covalent bonds, the necessary complement can be achieved by Lewis-acid-containing receptors, just the opposite to the way metal cations are coordinated to free electron pairs (ether, amine, carbonyl) of the receptors.

Since these concepts have been described in detail in previous volumes of this series and in other reviews [1], only some representative examples will be briefly presented here. We will concentrate more specifically on the binding of oxoanions, such as carboxylates and phosphates, since these are the most important anionic functional groups present in organic molecules.

2 Complexation of Carboxylic Acids and Carboxylates

The first attempts to develop receptor models for guests containing carboxylate groups concentrated on protonated macrocyclic oligoamines. These compounds effectively bind their guests via a combination of electrostatic interactions and hydrogen bonds. Kimura [2] and Lehn [3] independently showed that the carboxylate affinities of oligoammonium macrocycles, like the triprotonated form of [18]aneN$_6$ (1) and the octaprotonated form of [32]aneN$_8$ (2), are mostly governed by electrostatic interactions, so that the binding constants become higher as the number of protonated host nitrogen atoms increases. Macrocycles 3a and 3b, analogous to 1 but featuring a nonpolar middle part, were designed as "ditopic coreceptors" for the complexation of dianionic substrates [4]. Pronounced chain length dependence of the binding of the dicarboxylate guests has been observed [5]. Similarly, the tris(diphenylmethane) cryptand 4 binds the well fitting adipate more strongly than other α,ω-dicarboxylates, and even more terephthalate, due to additional arene–arene interac-

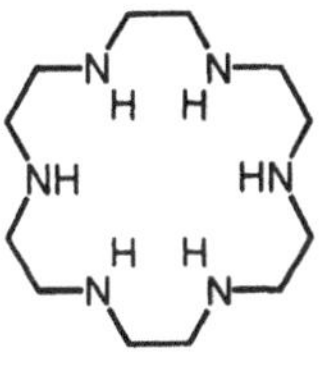

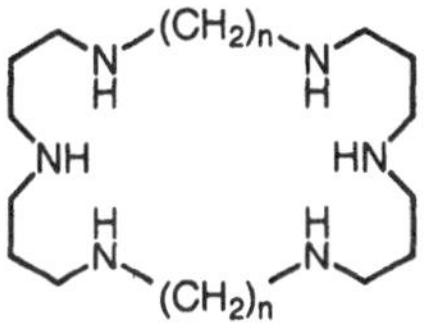

3a: n = 7
3b: n = 10

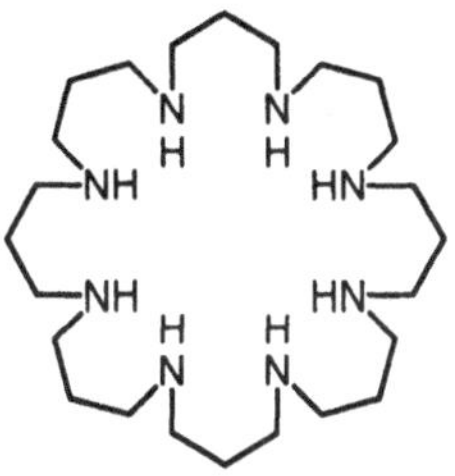

1   2

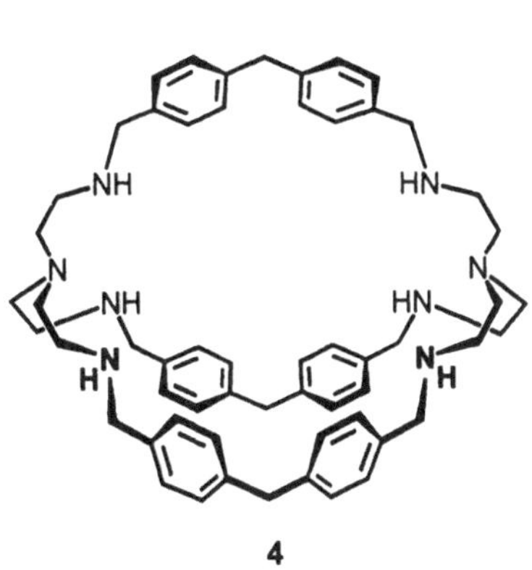

4

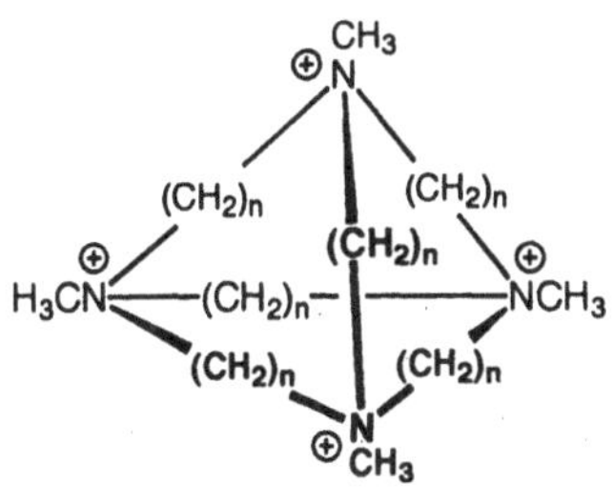

5a: n = 6
5b: n = 8

tions [6]. Comparable results where recently reported for a related water-soluble bisnaphthalene monocycle [7].

Tetrahedral tetraazonium compounds **5a** and **5b**, developed by Schmidt-chen, bind carboxylates, such as formiate, acetate, and benzoate, among a variety of other anions [8]. Although these receptors do not display hydrogen bond interactions, they are endowed with well localized polycationic centers, which are independent on the acidity of the medium. A ditopic receptor build up of both a **5a** and a **5b** subunits bridged by a *p*-xylylene spacer showed selectivity towards the largest members of a series of dianionic probes [9].

A major limitation of oligoamine receptors is the requirement of strongly acidic media to achieve their full protonation. This problem can be avoided by the use of more basic functional groups, like guanidines (pK_a 13.5). Furthermore, two well oriented hydrogen bonds can be formed between guanidinium cations and bidentate oxoanions under neutral conditions, ensuring both strong electrostatic interaction (ion pairing) and structural organization (H bonds). However, a comparative study of carboxylate and phosphate complexation by acyclic ammonium and guanidinium receptors revealed that the later, with a more delocalized positive charge, gave rise to weaker complexes, a fact that points out the importance of electrostatic interactions for anion binding [10]. Some examples of guanidinium-containing macrocycles (**6–8**) were described by Lehn. Comparison of binding constants of such receptors with those of acyclic analogues showed only moderate macrocyclic effects, although some selectivity was observed [11].

An example of the above mentioned "cascade complexation" of carboxylates by macrocyclic receptors containing metal ionic centers is the inclusion of oxalate by the "dien" dicobalt complex **9** (Martell, Mitsokaitis) [12]. Similarly, the β-cyclodextrin (β-CD) **10**, modified with a zinc cation bound by a triamine side chain, encapsulates anions like 1-adamantylcarboxylate in its cavity, fixing them by combined electrostatic and hydrophobic interactions [13]. Zinc's group achieved the enantioselective transport of the potassium salts of *N*-protected amino acids and dipeptides by making use of the cation affinity of

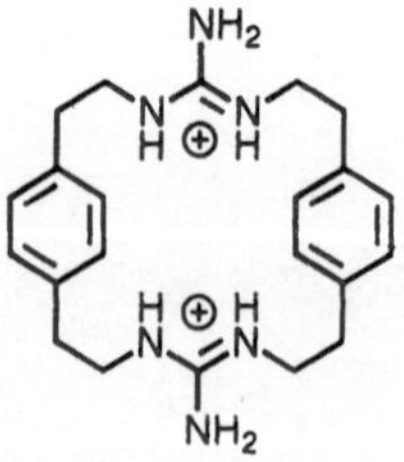

6 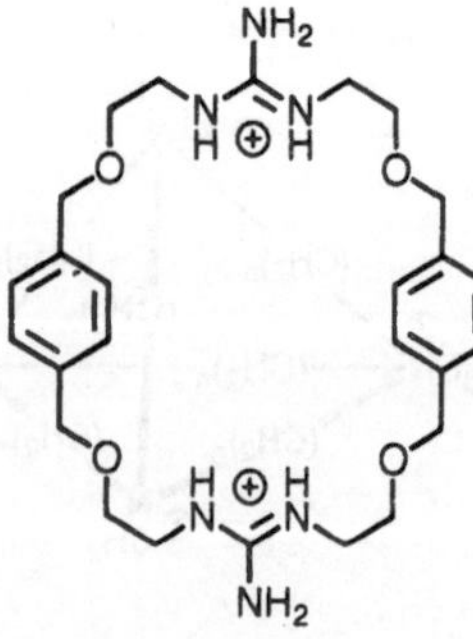   7 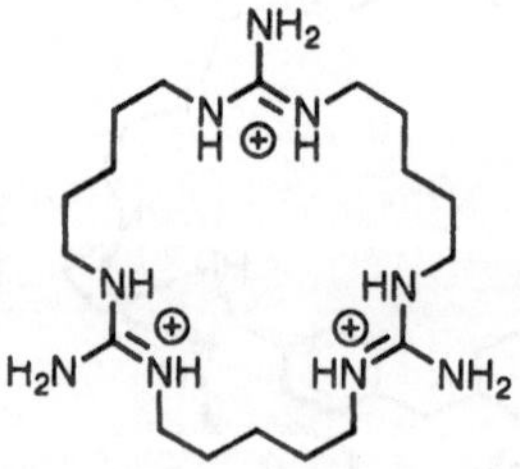   8

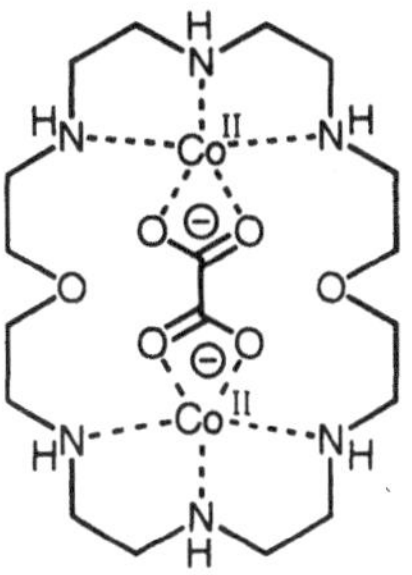

9 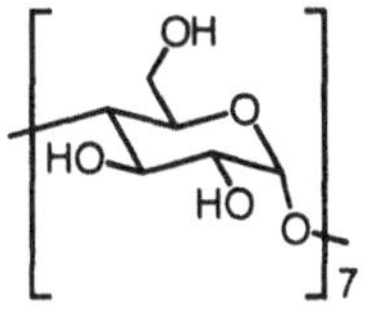   β-CD 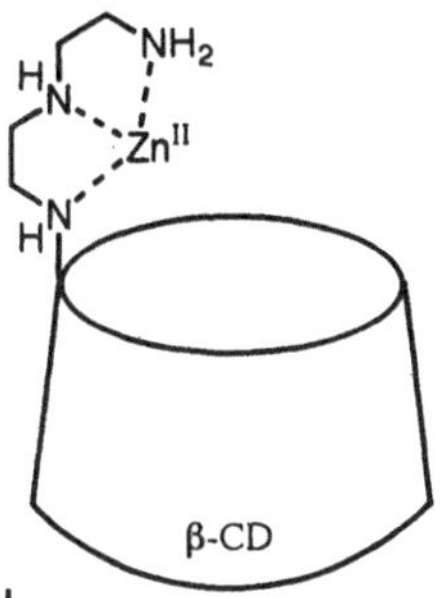   **10**

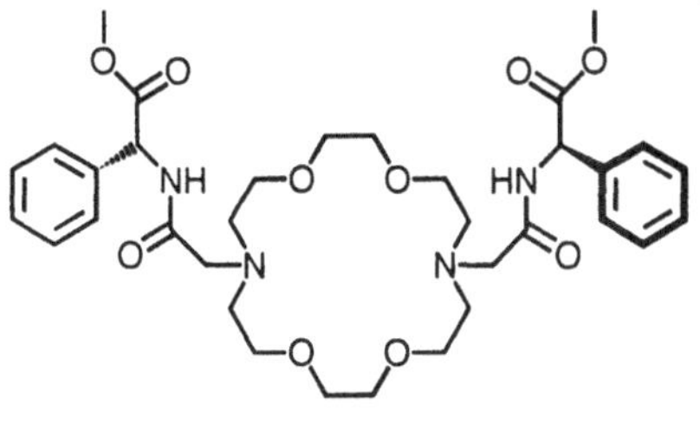

11

azacrown-derived lariat ethers like **11**, with the chiral information coming from D-2-phenylglycine units in the side chains [14]. A calix[4]arene modified with two cobalticinium units at the upper rim turned out to be an adipate binder [15].

The macrocyclic receptor models mentioned so far were designed for complexation in aqueous solutions, using highly charged binding sites in preorganized, sometimes rigid macrocyclic frameworks with the aim to overcome the unfavorable strong solvation of the guests. In contrast, most of the recently published works on this field focus on open-chain, often cleft- or tweezer-like systems. With the prize of abandoning utmost preorganization of the binding sites, the syntheses are in general easier and provide higher yields than the macro(oligo)cyclic receptors, which must be quite large to encapsulate organic substrates. The main molecular structure can also be modified easily in most cases to be adapted to different types of substrates like amino acids or nucleotides. Selectivity of molecular recognition of receptors having a rigid skeleton derives from the correct positioning in distance and the spatial direction of the binding functions. Other hosts, more flexible in nature, may nevertheless also exhibit distinct discrimination properties, the driving force being strong enthalpic interactions of their binding functions with complementary subunits of the guests. In these cases the substrate forces the receptor during the complexation process to assume a favorable conformation that fits the claims of the former.

Chiral bicyclic guanidinium receptors (e.g. **12–14**, S,S-configuration shown) have been developed in our group from aminoacid precursors [16]. Improved synthetic procedures for such compounds were later reported by Schmidtchen [17] (who in 1980 presented a first type of achiral bicycloguanidinium hosts [18] that form complexes with several oxoanions [19]). Most derivatives are highly lipophilic despite their ionic structure and freely soluble in common organic solvents but not in water. Therefore, extraction of organic oxoanions, such as carboxylates (see below for phosphates), from water is very efficient with **12a** [20]. Additional stacking interactions with the naphthoyl substituents are revealed by complexation-induced shifts (CIS) in NMR spectra and account for the selectivity observed for aromatic carboxylates over aliphatic ones. No such effects exhibited neither analogue **14**, endowed with aliphatic side arms, nor hosts **13a–c**. In the latter three cases, the shorter link between the sidearms and the core, which might inhibit favorable host-guest contacts, is thought to be the reason. A different phenomenon was observed for the disulfide **12c** and the diamine **12d**. Their side arms are more flexible than those of the diester **12a** and the diamide **12b** since no mesomeric effects of carbonyl groups hinder the single bond rotations. No CIS have been detected but upfield shifts of the signals of the aliphatic triazabicyclodecane protons. Probably, intramolecular cation-π attraction between the naphthyl rings and the guanidinium function is stronger than any intermolecular stacking with aromatic units of the counterions [21].

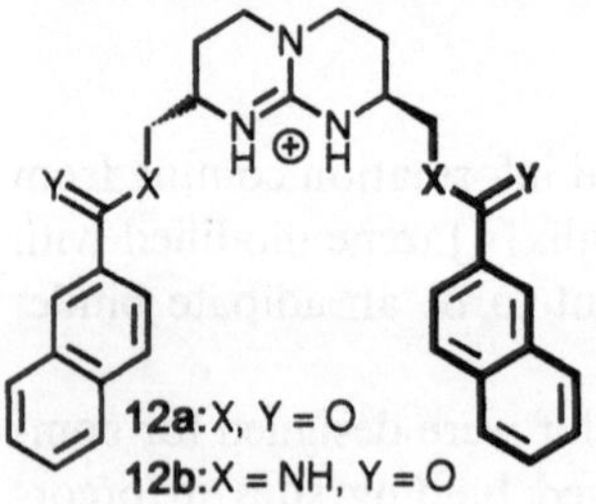

12a: X, Y = O
12b: X = NH, Y = O
12c: X = S, Y = H_2
12d: X = N(CH_2-2-Naphthyl), Y = H_2

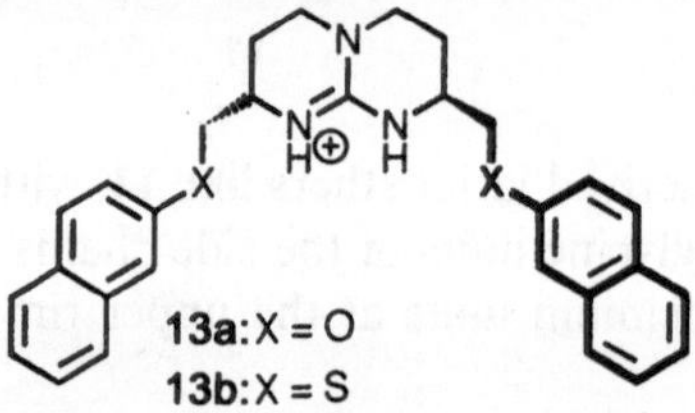

13a: X = O
13b: X = S
13c: X = NTos

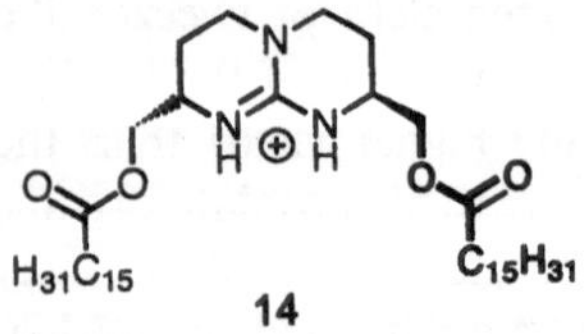

14

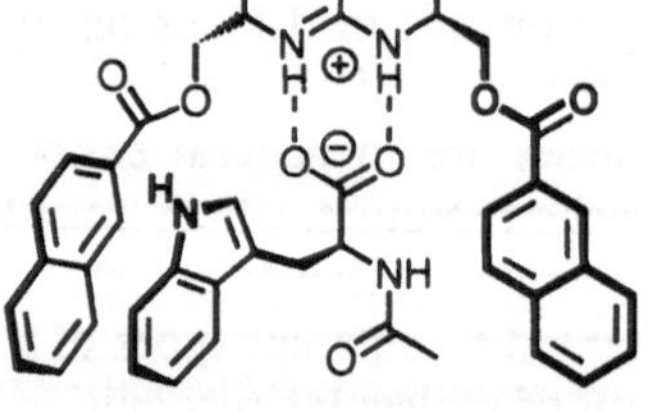

15

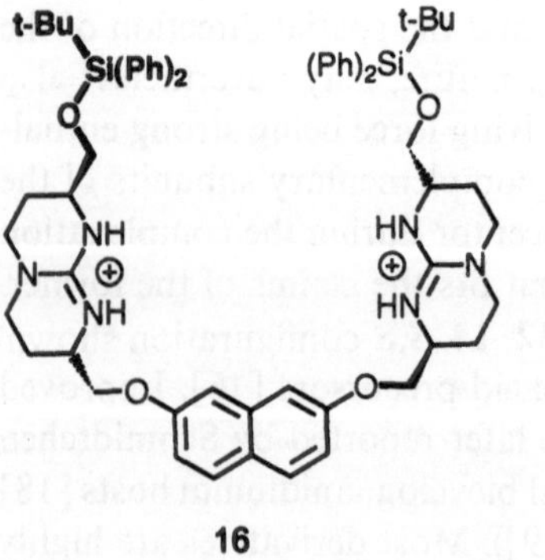

16

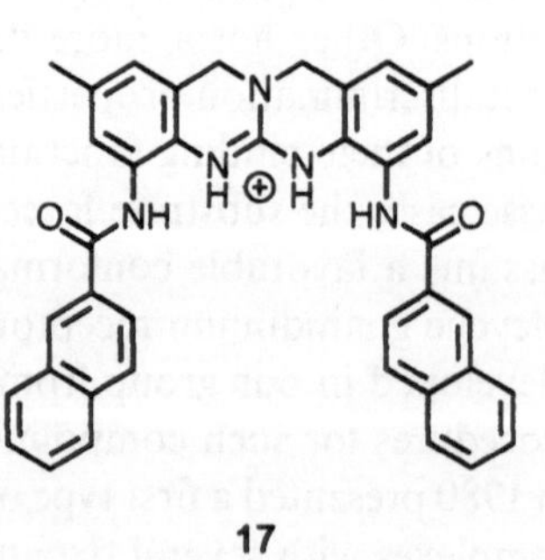

17

Although the chiral centers of **12a** (both the S,S and R,R enantiomers are accessible in optically pure forms) are not very close to the chiral centers of optically active carboxylates and the molecule is quite flexible, a moderate L-enantioselectivity was observed for N-acetyltryptophan extraction as first example of chiral recognition of anions [22]. The selectivity can be explained by the simultaneous action of both naphthoyl residues around the tight and highly structured guanidinium-carboxylate salt bridge, one interacting by stacking and the other providing some enantio-dependent hindrance to the acetamide group (compare formula **15**). According to this explanation, higher L-selectivity has been observed for the bulkier N-BOC-tryptophan.

With the aim of binding dicarboxylic anions, Schmidtchen and coworkers synthesized receptor **16** by means of connecting two bicycloguanidinium building blocks with a naphthalene spacer [23]. Due to its flexible framework it binds dianions that range in size from carbonate up to p-phenylenbis(3-acrylate) with maximum association for malonate and 3-nitroisophthalate ($K_{ass} = 16500$ M^{-1}, 14500 M^{-1}, respectively, in methanol) [24].

Recently, the synthesis of a new type of guanidinium-containing hosts has been achieved in our laboratories. The two annelated benzene rings of **17** cause a higher rigidity of the molecular skeleton and the four NH protons are ideally positioned for the formation of four hydrogen bonds with oxoanions. Although the backbone lacks the chirality provided by **12–16**, enantioselective complexation can be obtained by substitution of the naphthoyl groups by suitable chiral side arms, such as N-protected amino acids [25].

The molecular clefts designed by Rebek for the complexation of a great variety of organic molecules by hydrogen bonds and additional π-stacking interactions are among the most prominent open-chain but rigid hosts [26]. As representative examples, **18** and **19**, which are capable of solubilizing a series of otherwise insoluble aliphatic and aromatic dicarboxylic acids in CDCl$_3$, should be mentioned [27].

18   **19**

Morán et al. recently reported receptor **20** that contains two pyridazine-2,4-dione units as polar anchor groups anellated to an aromatic framework [28] as well as the xanthone compound **21** where one of the binding sites is a benzoylamido group [29]. Compound **20** fits the binding pattern required for malonic acids, whereas **21** forms complexes with aromatic carboxylic acids like p-ethoxybenzoic acid in CDCl$_3$.

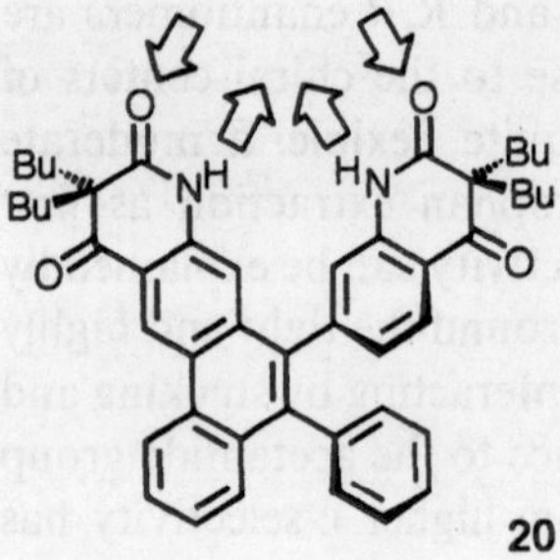

20 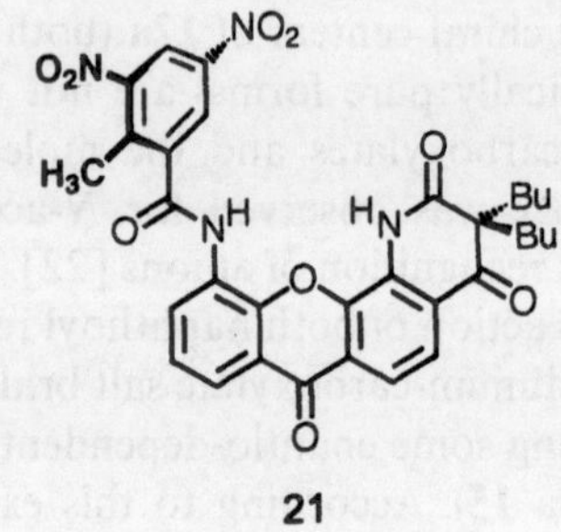   21

In the last few years, Hamilton has published reports of a series of open chain or cyclic receptors featuring 2-acylaminopyridine units as combined H-bond donor and acceptor groups [30]. The simple terephthaloyl derivative **22** forms stable complexes with dicarboxylic acids of appropriate chain length, like adipic and glutaric acid, in CDCl$_3$ [31]. Additionally, complexation influences the *s-cis:s-trans* ratio of the amide bond in acylprolines [32].

With the aim of increasing the strength of the host-guest interactions, the amidopyridyl groups of **22** have been replaced by urea, thiourea, and guanidinium units [33], thus yielding receptors **23** and **24** for carboxylate anions rather than for carboxylic acids [34]. Bis(amidopyridine) compound **25** and other analogues of **22** with different aromatic spacers have been object of investigation of controlled self-assembly [35] to well-ordered multicomponent complexes in solid state. "Overlong" non-fitting diacids do not form distorted 1:1 complexes with their counterparts but instead two- and three-dimensional networks, in which the two components are arranged alternately, at times linked to each neighboring molecule by two hydrogen bonds. For example, **25** and 1,8-octanedicarboxylic (sebacic) acid or 1,12-dodeanedicarboxylic acid form ribbon like structures in their cocrystal [36]. In other cases helical strands [37] or figure-of-eight-like 2:2 arrays have been observed [38].

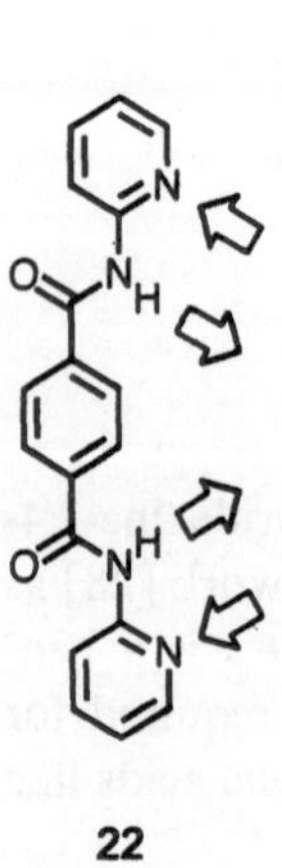

22 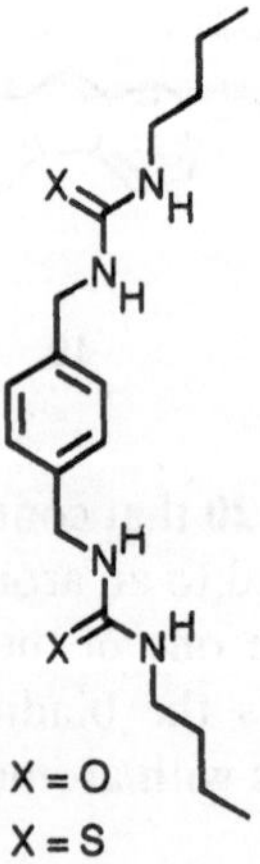   23a: X = O   24   25
                      23b: X = S

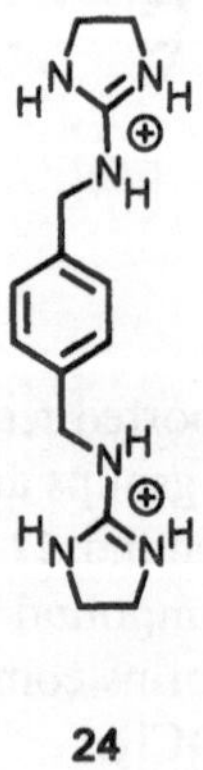

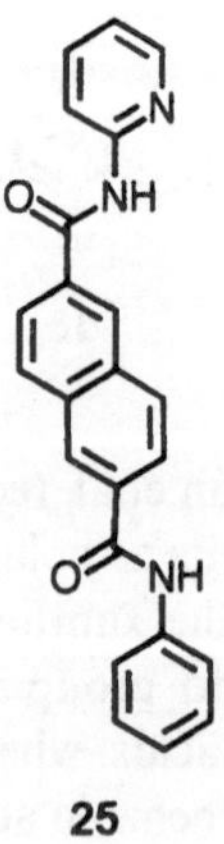

Diederich and coworkers attached two 2-amidopyridine units to 9,9'-spirobifluorene, 1,1'-binaphthyl, and heptahelicene building-blocks, so to obtain bis(bidentate) chiral molecular clefts [39, 40]. Hosts **26**, **27a**, **28**, and **29** complex dicarboxylic acids ranging in size from diethylmalonic to pimelic (heptanedioic) acid in CDCl$_3$. Furthermore, in contrast to the more flexible binaphthyl receptors **26** and **27a**, the rigid helical chiral framework of **28** causes distinct enantiodiscrimination in binding of N-protected aminoacids [41]. Some chiral recognition of tartaric acid derivatives has been observed by Hamilton et al. in independent studies with host **27b** [42].

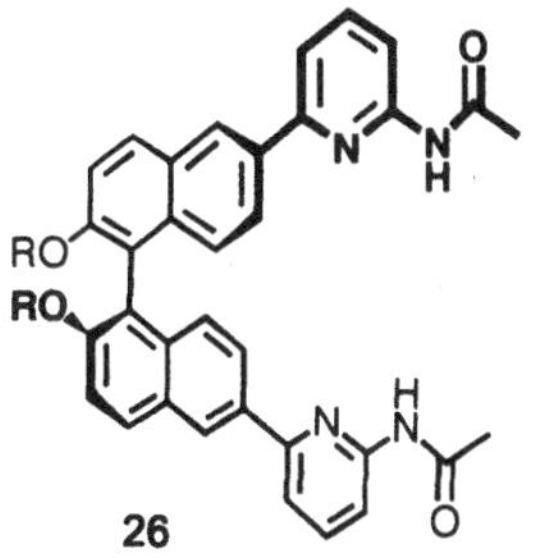

26

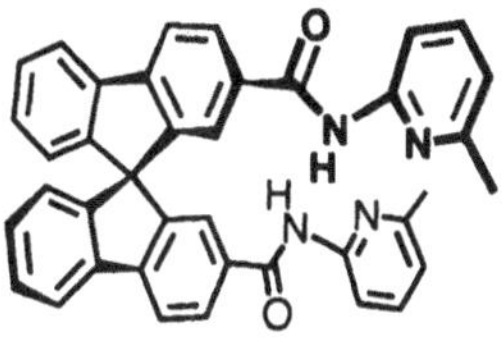

28

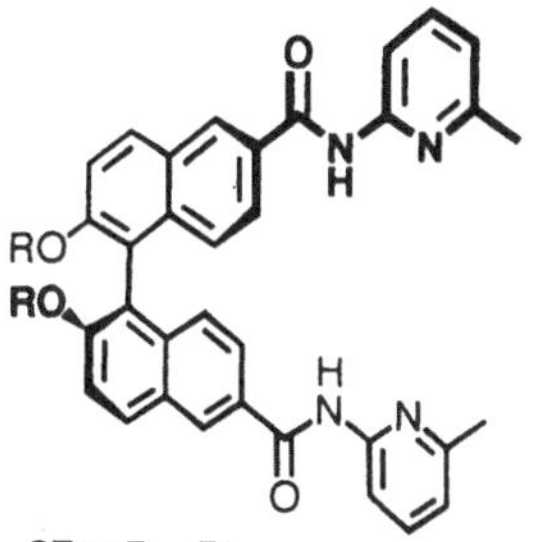

27a: R = Bn
27b: R = CH$_3$

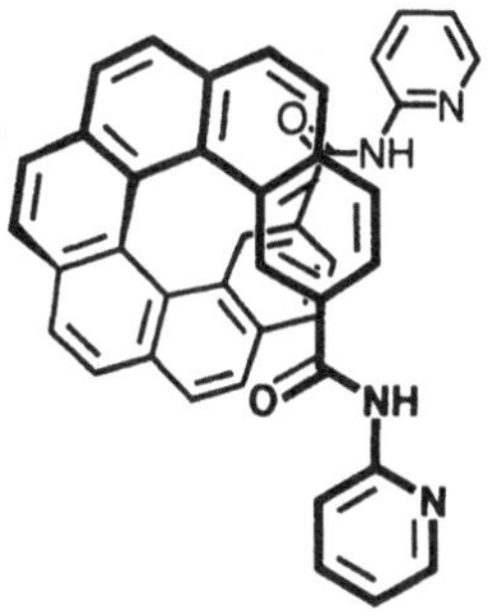

29

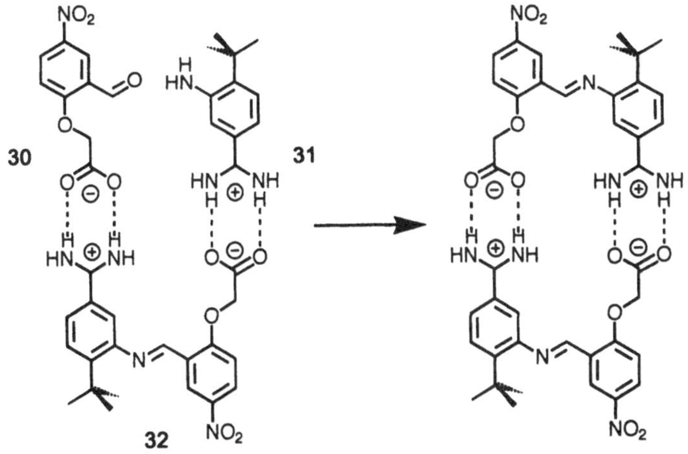

30

31

32

Artificial self-replicating systems have become of increasing interest recently [43]. Only one concept designed by von Kiedrowski will be briefly presented here since it is based on carboxylate-amidinium recognition [44]. The formation of a Schiff's base bond between formylphenoxyacetate **30** and aminobenzamidinium compound **31** is template-catalyzed by **32** as sketched out in the drawing. In a favorable conformation, the dimer-binding sites point in the same direction and preorganize both reaction partners upon formation of a ternary complex that is stabilized by two salt bridges with four hydrogen bonds.

3 Complexation of Amino Acids

Much interest in the field of molecular recognition of organic compounds has been focused on bio-relevant species, among which amino acids are an important class. Receptors ideally designed for selective structural and chiral recognition have to provide three different types of binding sites. The simultaneous recognition of the positively charged ammonium group, the anionic carboxylate function, and the side chain must be performed by a chiral three-point fixation. The selective binding of the side chain may turn out to be the most difficult task since many of the natural amino acids lack charged or polar functional groups in this part, and weak hydrophobic interactions or mere steric repulsions are the only means of discrimination. The zwitterionic nature amino acids display in aqueous solution represents another major problem. Due to the two charges, complexation in water is a difficult venture because strong solvation has to be outmatched. Moreover, a reduced ability of both ionic subunits to form complexes has to be taken in account owing to their mutual neighborhood [45]. This is why it is in general easier to complex either the protonated ammonium or the deprotonated carboxylate form (not to speak of *N*- or *C*-protected derivatives [46]). In a pioneering work, for example, Cram et al. reported extraction of perchlorate salts out of water into a $CD_3CN/CHCl_3$ phase by 1,1'-binaphthyl crowns [47].

Some authors based their approach to selective binding of the more lipophilic *a*-amino acids in water on hydrophobic effects using water-soluble, cavity-containing cyclophanes for the inclusion of only the apolar 'tail' under renouncement of any attractive interaction of the hosts with the zwitterionic 'head'. Kaifer and coworkers made use of the strong affinity of Stoddart's cyclobis(paraquat-*p*-phenylene) tetracation **33** for electron-rich aromatic substrates to achieve exclusive binding of some aromatic α-amino acids (Trp, Tyr) in acidic aqueous solution [48]. Aoyama et al. reported on selectivities of the calix[4]pyrogallolarene **34** with respect to chain length and π-basicity of aliphatic and aromatic amino acids, respectively [49]. Cyclodextrins are likewise water-soluble and provide a lipophilic interior. Tabushi modified β-cyclodextrin with a 1-pyrrolidinyl and a carboxyphenyl substituent to counterbalance the

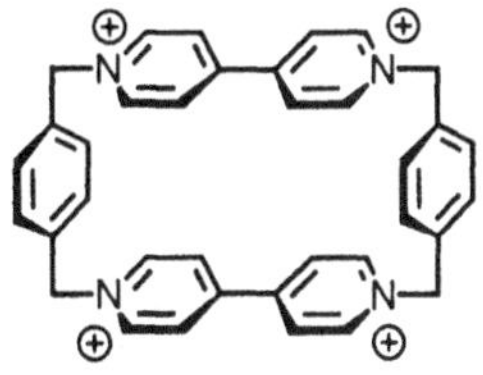

33

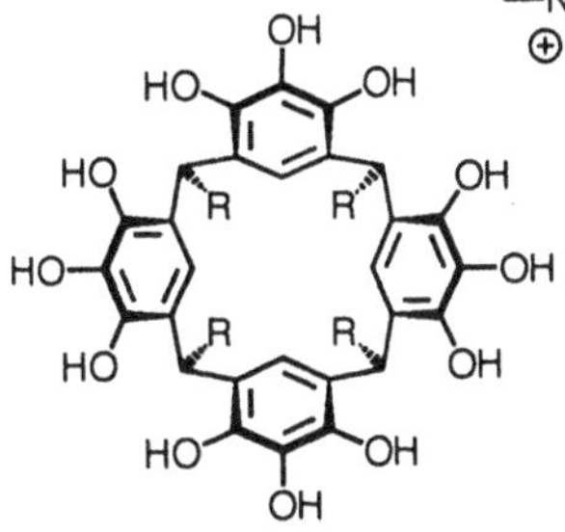

34: R = (CH$_2$)$_2$SO$_3$Na

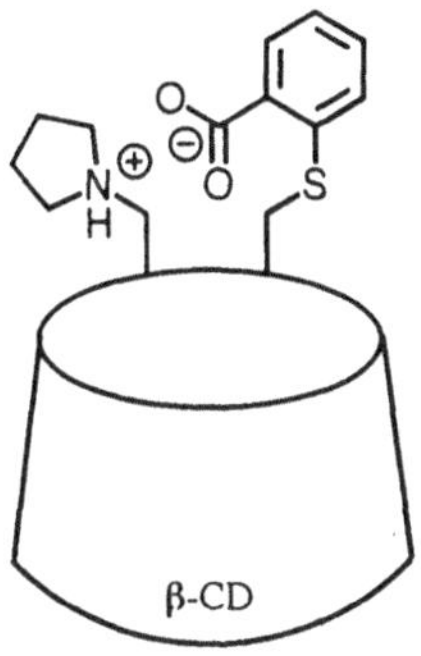

35

two charges of the guests and observed moderate complexation of tryptophan at pH 8.9 by **35** (only one regioisomer shown) [50].

In analogy to the carboxylate binding by zinc-containing cyclodextrin **10** (see Sect. 2), Lewis acidic centers such as a copper(II) histamine unit may also serve for the chelation of the (deprotonated) 2-aminoacetate substructure of α-amino acids [51]. Rizzarelli, Marchelli et al. used a respective β-cyclodextrin derivative for the formation of the ternary complexes **36** and **37** with racemic

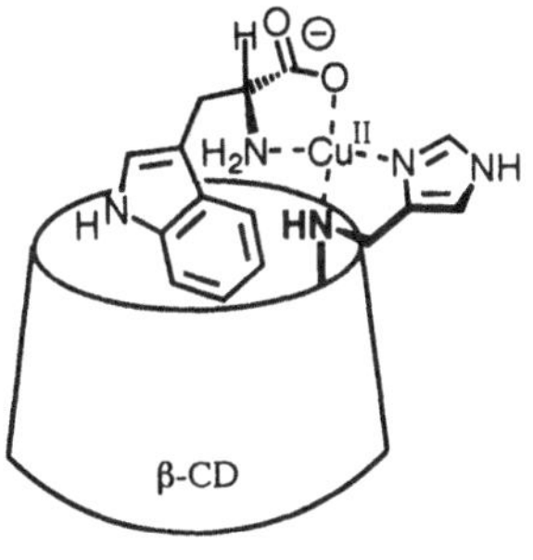

36

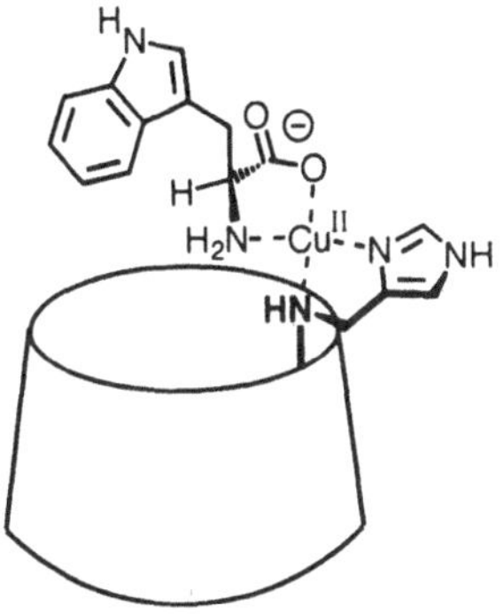

37

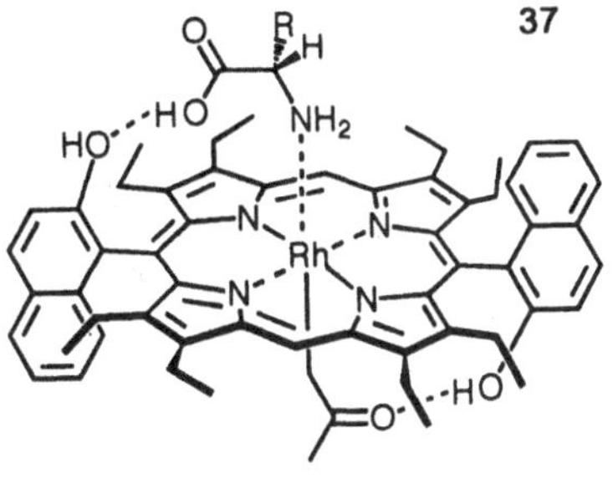

38

tryptophan. For reasons of stereochemistry, only in case of the D-enantiomer the indole ring can be engulfed by the host framework and HPLC-separation has been achieved with an achiral reversed-phase column.

According to Aoyama, Ogoshi and coworkers, a rhodioporphyrin providing a naphthol hydroxy group as a second docking point may be applied as ligand for lipophilic amino acids in chloroform [52]. The guests are bound by a salt bridge between the cation and the amino group, and additionally by a hydrogen bond between the naphthol subunit and the acid group as depicted in formula 38.

Recognition of amino acids has been attempted more frequently with receptors that are likewise zwitterionic in nature (compare 35). Some early studies include the transport of phenylalanine by a merocyanine dye through a liposomal bilayer [53]. Rebek's dicarboxylate-complexing cleft 9 (see Sect. 2) turned out to be a selective (though achiral) binder of trypthophane, phenylalanine, and tyrosine methyl ether [54]. A reasonable structure for a phenylalanine complex of 2 : 1 stoichiometry, as deduced from NMR studies, is schematically represented in formula 39 [55].

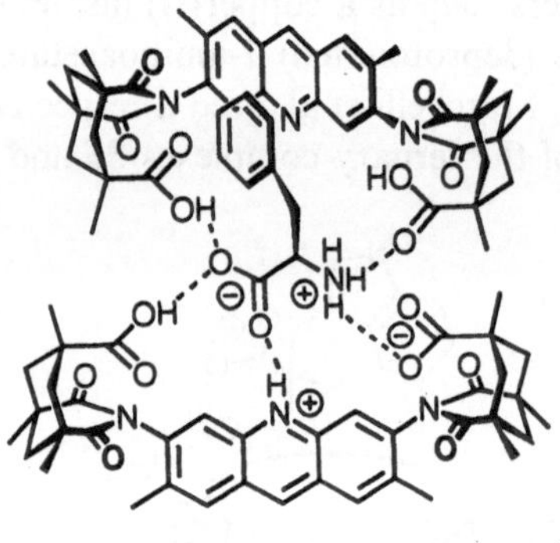

39

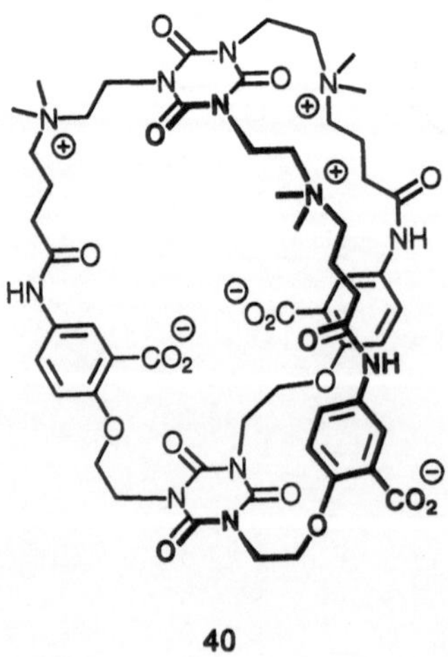

40

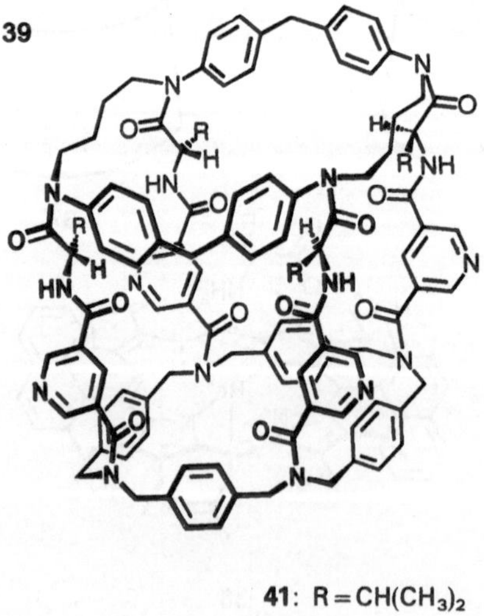

41: R = CH(CH₃)₂

Askew's tri(zwitterionic) cavity compound **40** strongly encapsulates 4-aminobutyric acid (GABA) in water but shows only little affinity towards nonfitting shorter or larger ω-amino acids [56]. Murakami et al. synthesized the huge oligocyclic cage molecule **41** introducing chiral information in form of L-valine building-blocks in the four bridges [57]. Selectively L-phenylalanine and D-tryptophan were attracted into the hydrophobic cavity in $D_2O/DMSO$ [58].

Following a different strategy for ω-amino acid hosts, Schmidtchen linearly coupled the binding sites for the two functional groups in receptor **42**: a tetrahedral tetraazonium unit as docking point for the carboxylate and an aza-crown for the ammonium group [59]. An important feature in the receptor design is that, contrary to the zwitterionic receptors described above, both binding sites are mutually non complementary, so auto-association that would hinder complexation is prevented. However, **42** showed no selectivity with respect to the chain length of the guests, and the simpler analogue **43** turned out to be a stronger binder of ω-amino acids in aqueous methanol.

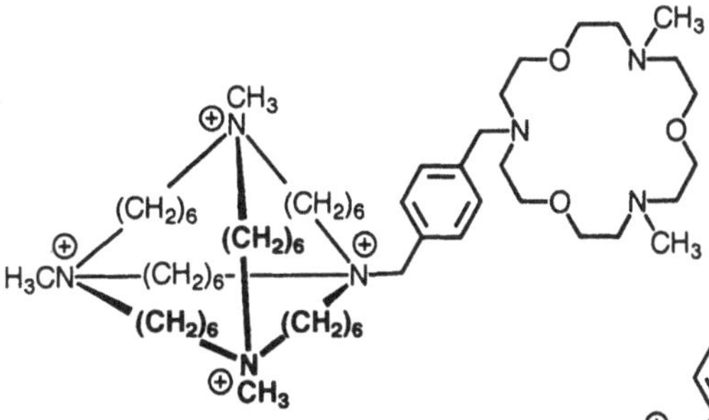

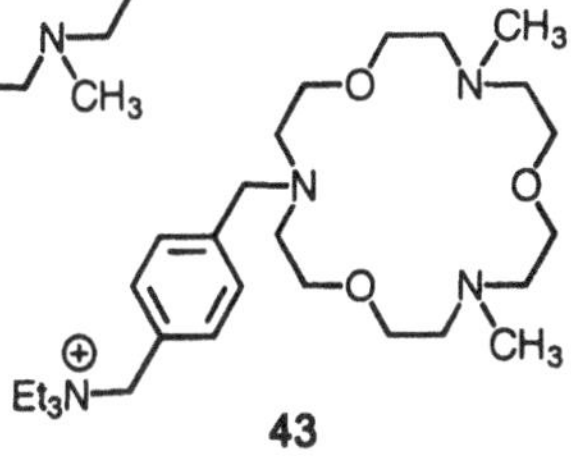

Refinement of the above strategy led our group to develop **44**, a tritopic receptor for the enantioselective recognition of aromatic α-amino acids under neutral conditions, featuring: (a) binding sites for carboxylate and ammonium in form of a cationic guanidinium function and a neutral crown ether, respectively, which are non-self-complementary and prevent the receptor from internal collapse or dimerization; (b) an aromatic planar surface (the naphthalene ring) for an additional stacking interaction with the side chain of aromatic amino acids; (c) a chiral structure [(S,S)-isomer shown] for enantioselective recognition [60]. The moderate chiral recognition observed for N-substituted tryptophan derivatives with receptor **12a** (see above) dramatically increases for *free* amino acids by replacing the mere steric interaction (one aromatic side chain) by a binding attraction (the azacrown). Thus, the (S,S) enantiomer removed ca. 40% of L-Trp or L-Phe from saturated water solutions into dichloromethane in single extraction experiments, whereas the corresponding D-enantiomers and aliphatic amino acids, like L-Val, were extracted much less efficiently. For example, no extraction of D-Trp or D-Phe was observed by NMR [reciprocally, only these isomers were extracted with the (R,R) receptor], and extraction of racemic amino acids yielded only minor amounts of D-Trp (0.5%) and D-Phe (2%), as determined by HPLC analysis of the corresponding diastereomeric L-Leu dipeptide derivatives (synthesized after the extraction to obtain separation) [61]. A competitive liquid-liquid single extraction of Phe, Trp, and Val mixtures led to a 100:97:6 ratio (NMR). The results of the extraction of a more complex mixture (13 amino acids) showed enhanced selectivity for Phe, although some lipophilic substrates, like Leu, were extracted significantly, too [60].

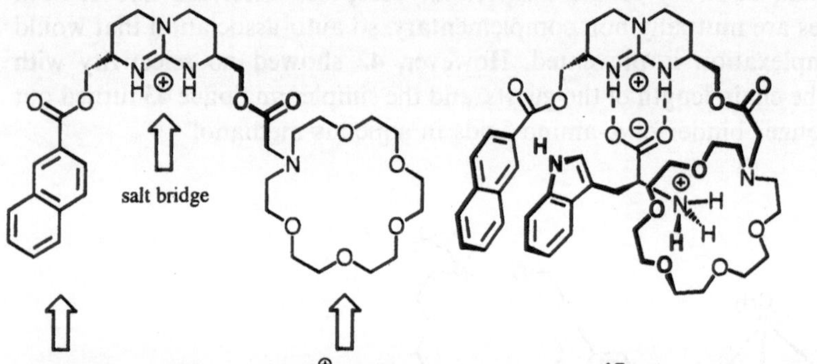

π-stacking   **44**   NH$_3^{\oplus}$-docking   **45**

Receptor model **44** is a flexible molecule, non-macrocyclic and without much conformational preorganization. Its selectivity has to be explained in terms of the almost perfect three point binding to the substrate (aromatic stacking of the amino acid side chain to the naphthoyl residue, ion bridging of the carboxylate with the guanidinium function, and ammonium binding by the crown). Molecular mechanics and molecular dynamics (500 ps at 300 K) calculations modeled a rather stable L-Trp complex **45** [62]. Several of the starting structures evolved toward partial dissociation of the interacting species through the progressive loss of two of the three attractive interactions. In some cases, however, the effect of the remaining anchor led to the rearrangement of the sidearms and their stable reassociation with the guest, fixing the host in a optimal conformation (Fig. 1). No such wrapping around the L-amino acids can be achieved with the (R,R)-enantiomer.

Analysis of the individual contributions of each building block to the binding enthalpy in a typical minimized structure reveals that, in the absence of solvent, almost half of the stabilization energy comes from the essential electrostatic interaction and hydrogen bonding of the bicyclic guanidinium with the carboxylate group. One additional third part is provided by the ammonium-crown interaction, and about one sixth arises from π-stacking between the naphthoyl ester and the indole ring. Thus, loss of complementarity for one of the

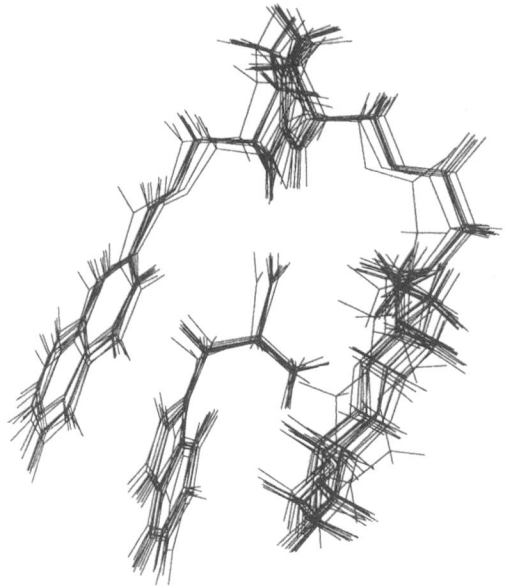

Fig. 1. Superposition of nine energy-minimized average structures of the L-Trp complex **45** of (*S,S*)-**44**

anchoring units (as in the D-enantiomer complex) can bring forth a substantial decrease in the stability of the complex, accounting for the pronounced enantioselectivity found experimentally. Whereas the binding of the zwitterionic part mainly determines the overall binding strength (some 85% for **45**), the third attractive force accounts for the molecular discrimination between different types of α-amino acids, and here is where future host generations can be tuned for new selectivities.

4 Complexation of Nucleotides

Recognition of nucleotides at the molecular level constitutes a recently very dynamic branch of supramolecular chemistry. Since nucleotides are built up by three structurally different subunits, namely the inorganic anionic side chain, the sugar, and the nucleobase, optimally designed receptors exhibiting not only strong binding but also pronounced selectivity should contain more than just the binding site for ionic interactions. It is decisive to distinguish the different nucleobases, which might proceed either via π-stacking or hydrogen-bonded base pairing. Important knowledge was gained by developing host molecules for derivatives of the heterobases [63], and additional contributions came from experiences made with ligands for phosphates and phosphoesters [64].

As in Sect. 2, the early works in this field will only be briefly summarized. Highly charged polyazoniamacro(oligo)cycles and guanidinium crowns bind phosphates and nucleotides via electrostatic interactions [1, 65], and polyammonium cations, like spermine, are generally known to associate with the phosphate backbone of RNA and DNA strands [66].

To come to a water-soluble, ditopic receptor model, Lehn et al. attached N_6-[24]crown-8 with an acridine-derived DNA-intercalator, so combining the anion binding strength of the first (in protonated state) [67] with the π-stacking capacity of the latter. Indeed, increased binding of ATP by 46 was observed as compared with the unsubstituted azacrown [68]. Additionally, the multifunctional host 46 catalyzes the hydrolysis of ATP [69] with a higher selectivity over ADP as compared with the parent crown but with a somewhat reduced effectivity. The reaction intermediate was shown to be the N-phosphorylated crown.

46

A different strategy for binding of nucleotides in water [70] is to apply hydrophilic cyclophanes that provide cavities for the encapsulation of the heteroaromatic bases of the guests. Lehn et al. synthesized the cavity containing "bis(intercalands)" 47 and 48 as well as their open chain analogues 49 and 50 [71]. In binding studies the latter two and not the more rigid macrobicycles yielded the best complexation. Still, neither a preference in binding of nucleotides relative to their parent nucleosides was observed nor discrimination of AMP versus the higher charged ADP and ATP.

Schneider and coworkers investigated the capability of the tetraazonia cyclophane 51 to bind nucleosides and nucleotides in water [72]. Its two diphenylmethane units border a lipophilic cavity for the uptake of likewise lipophilic molecules or subunits. As evidenced by ^{1}H NMR studies of the complexes, only in the case of the purine derived nucleosides and nucleotides (A, G, AXP, GMP) the base moiety is included inside the host cavity but not in the case of the pyrimidine analogues (U, C, UMP, CMP, TMP). Due to their hydrophilic nature, the sugar and phosphate groups remain outside the niche.

One of the motive powers for the design of host molecules for nucleotides is their potential utility in chemotherapy. A number of nucleotide analogues exhibit antiviral activity in vitro [73]; for example, 2',3'-dideoxynucleosides are potent chain-terminating inhibitors of HIV reverse transcripase [74]. However, charged and hydrophilic as they are, they can hardly penetrate across cell

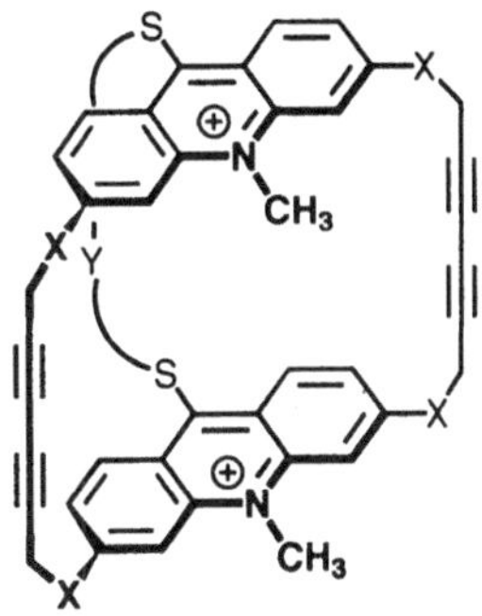

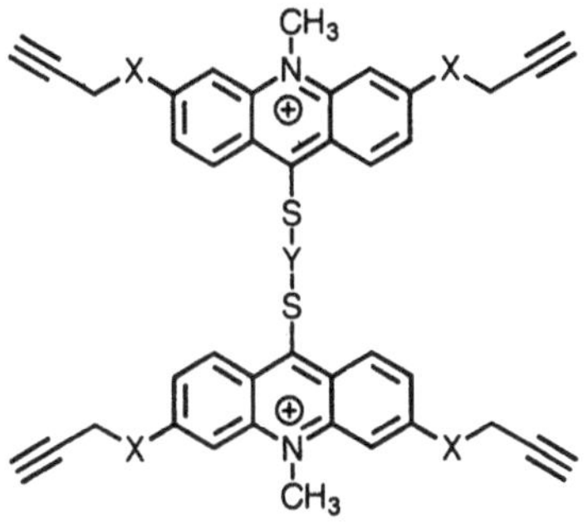

47: X = O
48: X = NH

47a, 49a: Y = $(CH_2)_6$
47b, 48, 49b, 50: Y = $(CH_2)_2O(CH_2)_2$
47c, 49c: Y = $(CH_2)_2O(CH_2)_2O(CH_2)_2$

49: X = O
50: X = NH

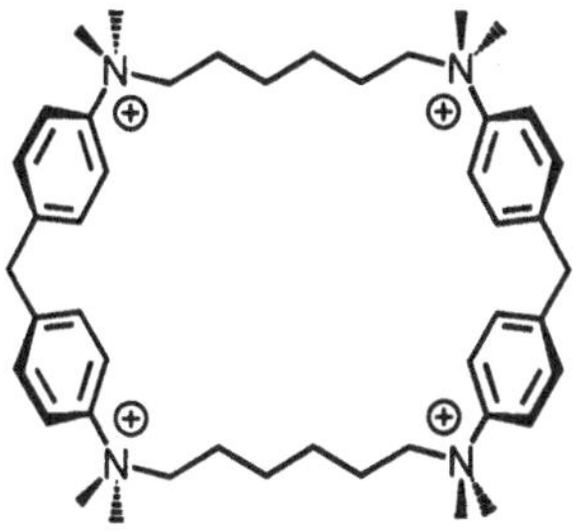

51

membranes and are often inactive in vivo. Artificial lipophilic carriers capable of enhancing the cellular uptake might be useful to augment the medical efficacy of such agents.

Around 1980, Tabushi and coworkers used the lipophilized, DABCO-derived diammonium salt **52** as a phase transfer reagent for the transport of nucleotides in three-phase experiments (H_2O–$CHCl_3$–H_2O) [75]. Whereas AMP was discriminated in a high degree, for ADP and ATP the acceleration rates were quite similar. The transport rates were significantly diminished for uracil and guanine nucleotides because of the low solubility of the resulting complexes.

Recently Diederich et al. tried to overcome the solubility problems by using modifications where the linear stearyl side chains are exchanged by branched ones [76]. Dionium salt **53** turned out to be an effective transport catalyst at physiological pH for all investigated compounds [AMP, CTP, 2',3'-dideoxy-TTP (ddTTP), and 3'-azido-dTTP (AZTTP)] and significantly improved the rates achieved with **52**. A chloroform solution of **53** extracts half an equivalent of ATP^{4-}, which indicates the formation of a neutral 2:1 complex. Analogously, bis(DABCO) tetracation **54** binds to one ATP molecule. However, the transport acceleration is about one order of magnitude smaller than that of **53** [77].

52

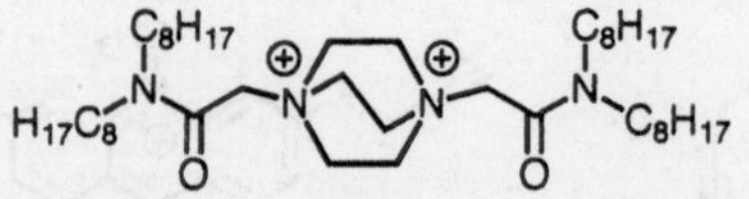

53

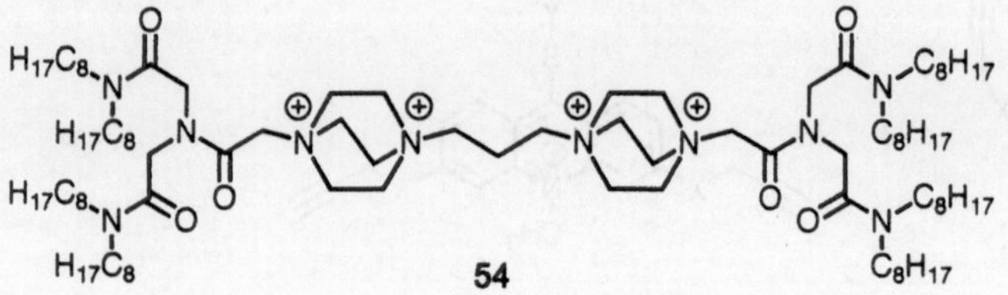

54

In a simple design of a ditopic receptor, Sessler et al. connected aliphatic amines as electrostatic binding sites with cytosine to obtain GMP-selectivity in binding [78]. In DMSO as solvent, the formation of a 1 : 1 complex of **55** and GMP free acid (H_2GMP) has been observed. The cytosine moiety stabilizes the complex by base pairing with the guanine ring additionally to the salt bridges between the ionic centers of the binding partners.

Sapphyrin (Sap) **56a**, a pentapyrrolic expanded porphyrin, was used by the same group in further attempts to apply large lipophilic cations as mediators for the carriage of anions through an organic membrane in classical U-tube experiments [79]. The unsubstituted parent compound **56a** needs an acidic source and a basic receiving phase to be effective in transporting monophosphates like AMP and GMP. This pH-dependency was strongly diminished by modifying sapphyrin with cytosine-containing side chains as in **56b** and **56c** [80]. However, a reversed order in substrate transport has been observed (GMP > AMP > CMP) for reason of the base pairing pattern of cytosine [81]. In further experiments, GMP transport was mediated by cooperative interaction of rubyrin, a hexapyrrolic expanded porphyrin, and lipophilic cytidine derivatives [82].

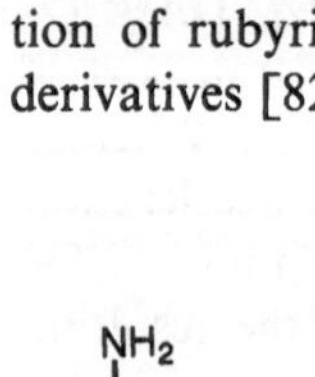

55

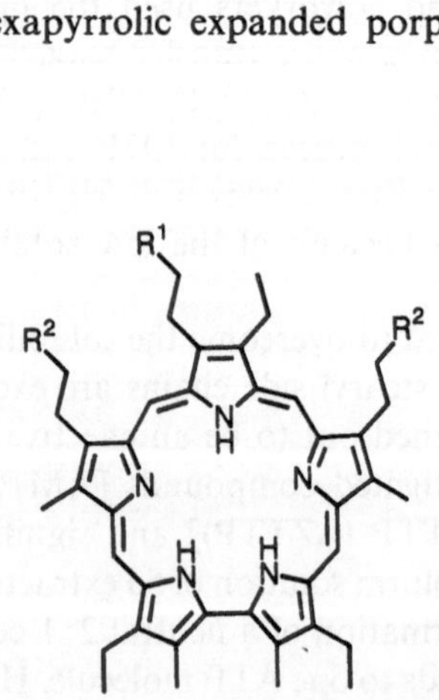

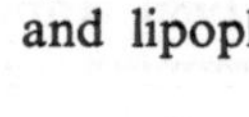

56a: R^1, R^2 = H
56b: R^1 = X, R^2 = H
56c: R^1 = H, R^2 = X

Koga et al. complexed oxo acids like methyl phenylphosphonate with bis(resorcinol) quinoline derivative **57** [83]. Anslyn and coworkers presented a more rigid polyazacleft containing hydrogen bond acceptor and donor sites in form of pyridine rings and amino groups, respectively. The formation of the 3:1 complex **58** with dibenzyl phosphate is assumed, in which the four components are spatially fixed by a net of H-bonds [84].

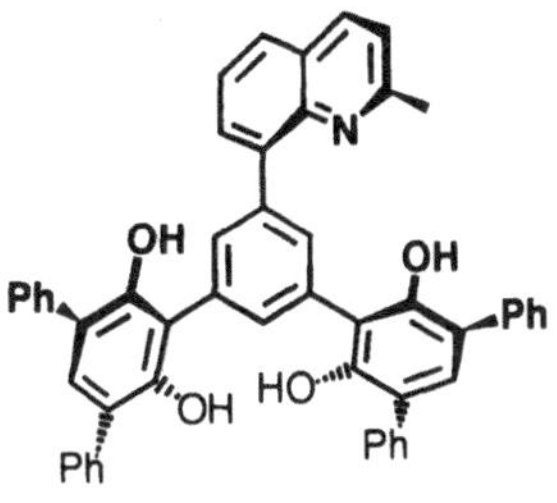

57 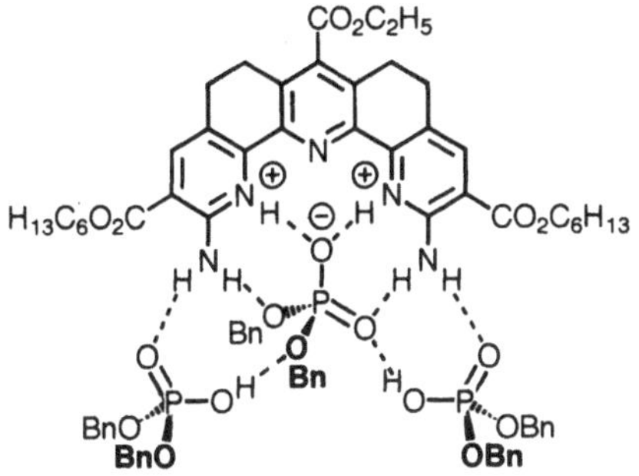   58

To achieve molecular recognition [85] and cleavage of phosphodiesters, Hamilton's group slightly altered the concept of directed hydrogen-bonding patterns for the complexation of carboxylic acids (see Sect. 2). The host molecule **59** is rigidized by intramolecular H bonds so that four guanidinum protons are directed to the interior of the cleft. The binding sites are not only well positioned to strongly bind diphenyl phosphate in CD_3CN [86], they may also stabilize trigonal-bipyramidal intermediates [87] which are formed in the course of phosphoester hydrolysis [88]. Thus, an excess of **59** in the presence of lutidine as a general base markedly increased the rate constant of the cleavage of phosphoester salts in acetonitrile.

With the goal of mimicking the catalytic center of staphylococcal nuclease (SNase), an essential feature of which are the guanidinum groups of two arginines [89], Anslyn designed the cleft **60**, as kind of hybrid of the azacleft **58** and the bis(guanidinium) compound **59** [90]. It exhibits high catalytic activity in transesterification and cleavage of aqueous mRNA under physiological conditions in the presence of an excess of imidazole as a general base. The mechanism of the formation of the trigonal-bipyramidal intermediate [91] is believed to be as schemed in formula **61** [92]. A future combination of the phosphate binding functions and the general base in one molecule promises to improve the catalysis [93].

The catalytic groups in the reactive center of ribonuclease A, a further polynucleotide-cleaving enzyme, are an imidazole (acting as a general base) and an imidazolinium cation (acting as a general acid). In a series of papers starting as early as 1978, Breslow and coworkers describe the use of β-cyclodextrinyl-bis(imidazoles) **62** as model compounds for RNase A [94]. In the course of the

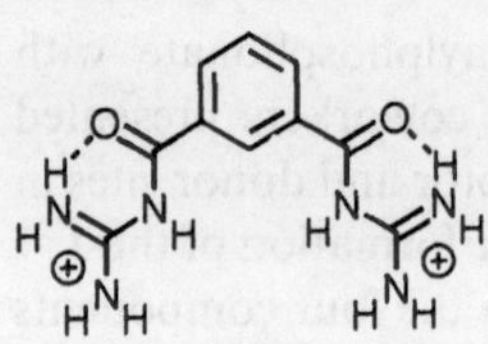

59

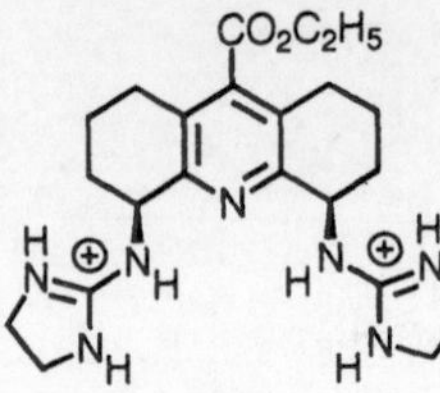

60

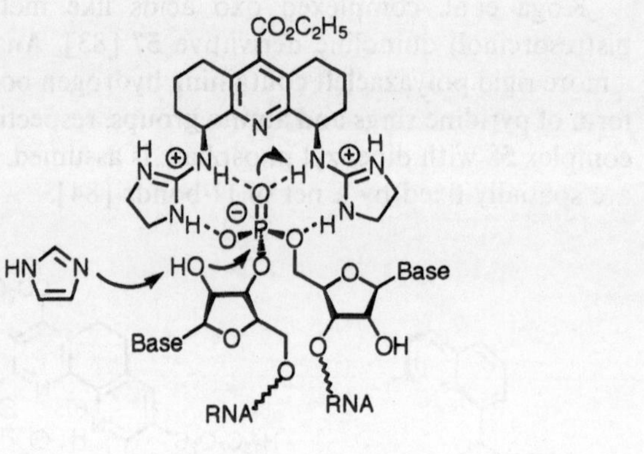

61

investigation it was shown that the 6A,6B-substituted isomer (the sugar rings are lettered A to G) with two vicinal sidearms is the most active in catalyzing the hydrolysis of the cyclic phosphate of 4-*tert*-butyl catechol **63** as substrate analogue. This observation stands in contrast to the so far favored 'in-line' mechanism of the enzyme with the reaction partners being 180° apart as is possible in the 6A,6C and 6A,6D regioisomers of **62** [95].

The general affinity of guanidinium cations for oxoanions and the spatial direction of the two NH-protons makes the bicycloguanidinium core presented in the preceding chapters also a useful building block for the design of nucleotide receptors [96]. Besides carboxylates (see Sect. 2), bis(naphthoyl) host **12a** binds several phophoesters in chloroform, for example, 1,1'-binaphthyl-2,2'-diyl phosphate; the complex of the (S)-enantiomer is shown in formula **64**. Still, despite the chiral nature of **12a**, no enatioselectivity was observed [97].

Additionally, **12a** showed affinities to adenosine monophosphates in liquid–liquid extraction experiments (D$_2$O/CDCl$_3$). The degrees of saturation, however, were mainly correlated with the relative lipophilicity of the nucleotides. The amounts of 3'-AMP$^-$, 2',3'-, and 3',5'-cAMP$^-$ (8-fold excesses) extrac-

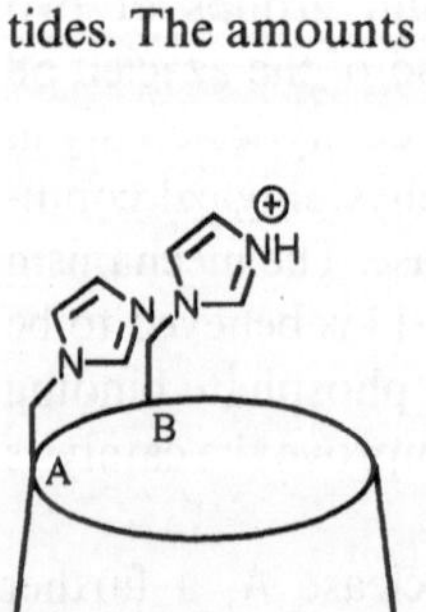

62 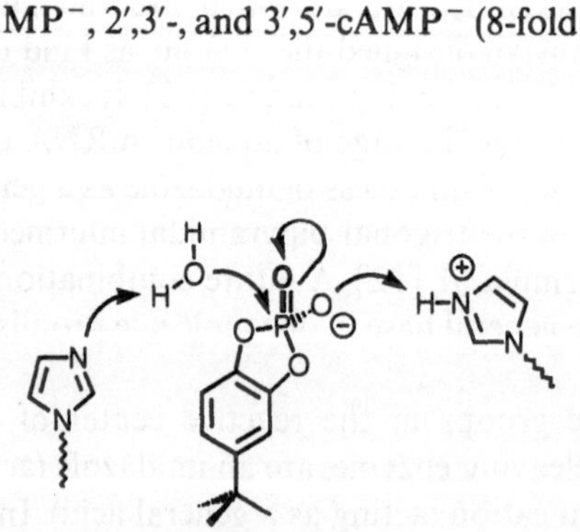   **63**

ted have been determined as 0.02, 0.2, and 0.5 equivalents, respectively, and as 0, 0.1, and 0.3 equivalents, respectively, in case of the aliphatic analogue **14** (Sect. 2) lacking aromatic side chains for stacking interactions.

In ditopic hosts **65** the two cationic substructures are linked by a naphthylene diether. Schmidtchen observed the formation of 1:1 phosphoester complexes in organic solution with disilylether **65a** and in water with diol **65b**. The flexible skeleton enables the hosts to wrap around phosphate forming two pairs of hydrogen bonds with perpendicular main planes as schemed in formula **66** for the 5'-TMP complex [98]. Similarly, the complexation of AMP derivatives, NAD, and other phosphates by the related host **16** (Sect. 2) and its desilylated modification have been reported recently [99].

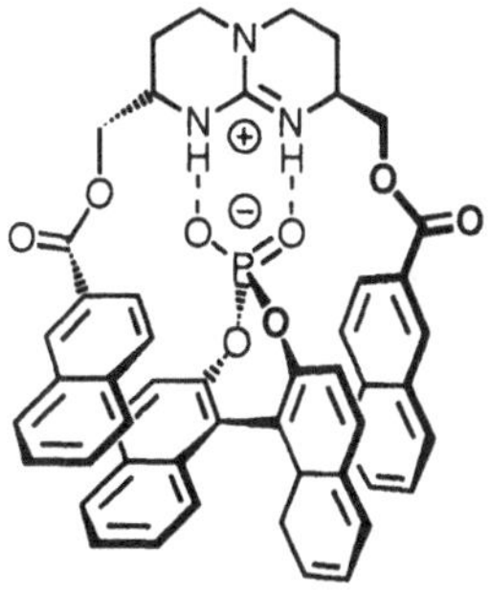

64

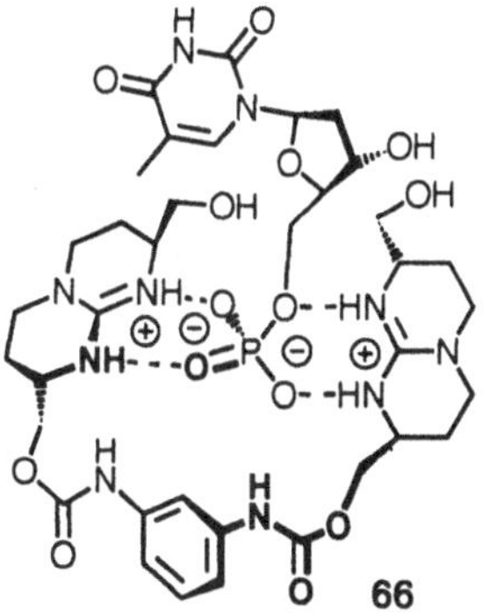

66

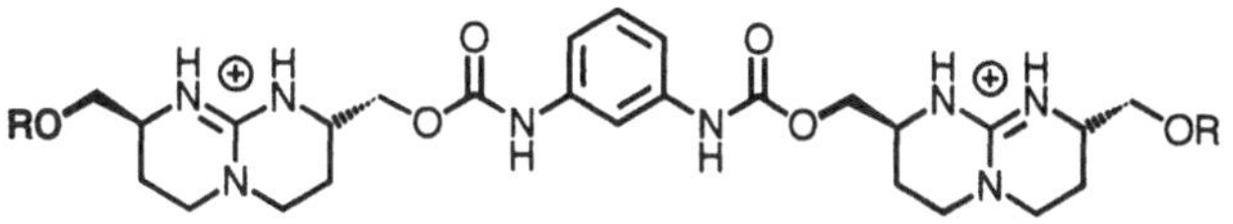

65a: R = SiPh$_2$Bu^t
65b: R = H

Compound **12a** was modified in our laboratories by exchanging one of the naphthoyl side chains for an uracil unit as a Hoogsteen-counterpart for the base pairing with adenine in a first attempt to transform the guanidinium compounds from poorly selective phosphoester-binders into nucleobase-distinguishing nucleotide receptors. Yet, the reduced lipophilicity of **67** prevented its application in extraction experiments. NMR shifts of the signals of the 3'-AMP^{2-} complex **68** detected in DMSO-d$_6$ confirmed the intended three-point host-guest docking (salt bridge, base paring, π-stacking). Downfield shifts of the guanidinium and uracil NH protons indicate the hydrogen bonding, and upfield shifts of all aromatic protons corroborate the arene–arene interactions [100].

In collaboration with the Rebek group, we designed host **69** in the search of a lipophilic receptor model that combines high adenine selectivity with the

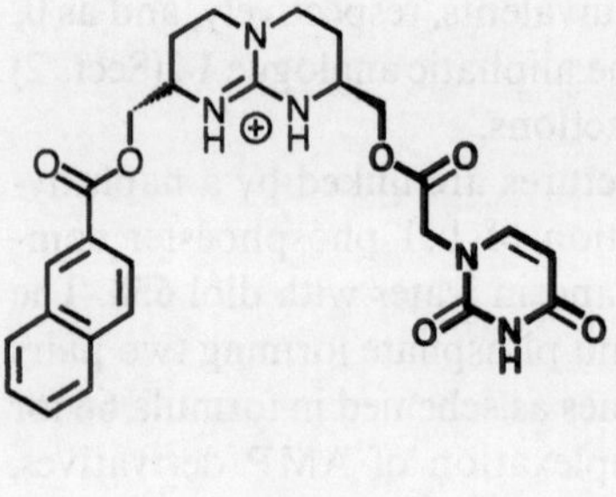

67 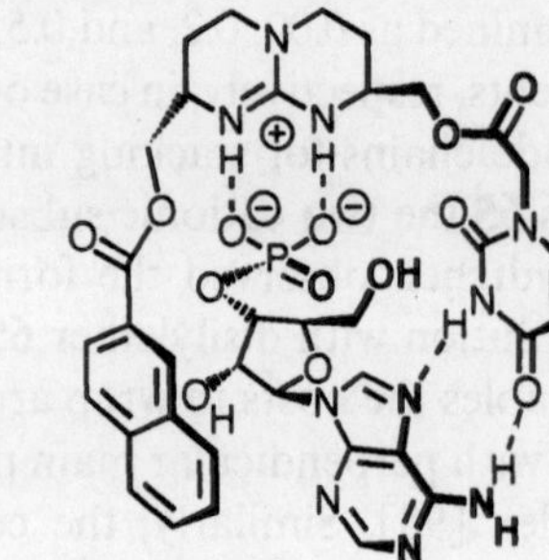   **68**

phosphate binding strength of guanidinium and evades any solubility problems. A molecular cleft built up from two Kemp's triacid derivatives linked together by a carbazole unit acts as nucleobase binding site. The electron-rich spacer renders possible π-stacking interactions with the heterobases and positions the converging imide functions in an adequate distance for simultaneous Watson-Crick and Hoogsteen-base pairing [101]. Liquid–liquid single extraction experiments, in which tenfold excessive aqueous nucleotide solutions were extracted with CH_2Cl_2 solutions of **69**, revealed a moderate preference for cAMP derivatives. One equivalent of both 2′,3′- and 3′,5′-cAMP was extracted into the organic phase, which indicates the formation of 1:1 complexes with multiple host-guest docking as schemed in formula **70**. For the cGMP analogues the extraction rates were 0.6 and 0.34, respectively, supposedly as a consequence of the 'wobble' contacts with less favorable hydrogen bonding of the guanosine nucleous inside the bis(imide) niche. The combination of salt bridging and nucleobase chelation as provided by receptor **69** is essential for binding strength and selectivity. This becomes manifest in the inferior data of the bisnaphthoyl and bispalmitoyl guanidinium hosts **12a** and **14** as mentioned above and the fact

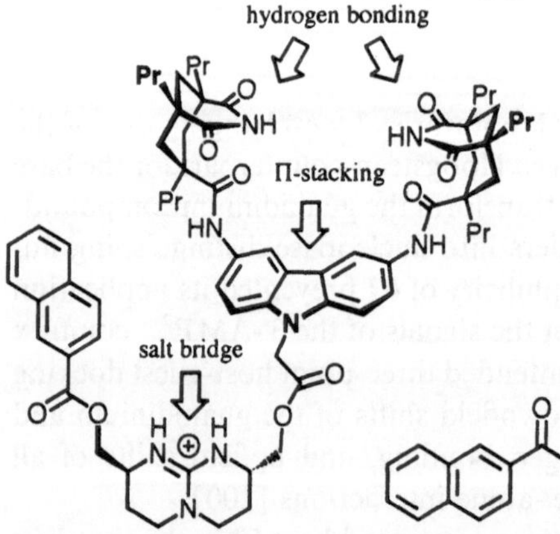

69 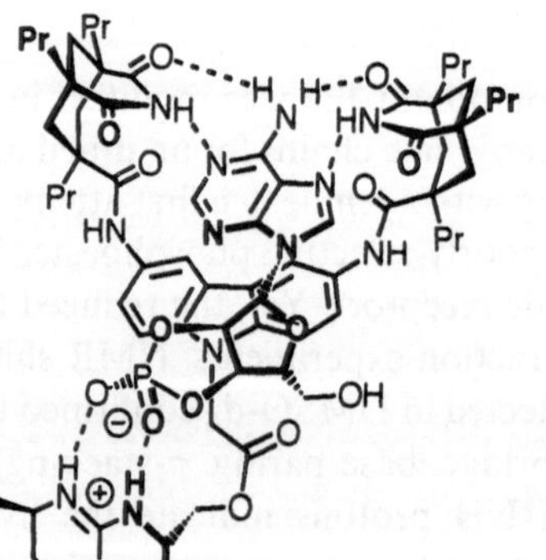   **70**

that the molecular cleft alone, without the guanidinium module, fails to extract significant amounts of nucleotides. Correspondingly, of all comparison host compounds only **69** turned out to be an efficient carrier for the exclusive transport of adenosine nucleotides through a 1,2-dichloroethane layer (U-shaped tube; 2′,3′-cAMP > 3′,5′-cAMP ≫ 3′-AMP, no transport of 2′,3′- and 3′,5′-cGMP) [102].

Receptor models **71a** and **71b** for dinucleotides (dinucleoside monophosphates) have been developed out of **69** by replacing the naphthoyl residue, which is not involved in any host-guest interactions in the complexes, by a second carbazolyl cleft. The adenine selectivity of the two 'pincers' was reflected in single extraction experiments: dissolved in CH_2Cl_2, **71a** removes one full equivalent of both adenylyl(3′ → 5′)adenosine (ApA) and 2′-deoxyadenylyl(3′ → 5′)-2′-deoxyadenosine [d(ApA)] **72** out of aqueous solutions of guest (8-fold excess) but only half the amount of d(ApG) [103].

Since in $CDCl_3$ the signals were broadened, NMR studies of the d(ApA) complex **73** were run in DMSO-d_6 where rapid association-dissociation equilibria exist. This solvent, however, weakens the hydrogen bonding interactions, and NOE's were observed from imide-NH to H8 of the adenines but not to H2. This corresponds with the base pairing pattern being mainly of the Hoogsteen type. Further NOE's indicated a adenine-sugar *syn* conformation of the 3′-free nucleoside (contact between NCH_2CO and H4′) and an *anti* conformation of the other (contact between NCH_2CO and H2′).

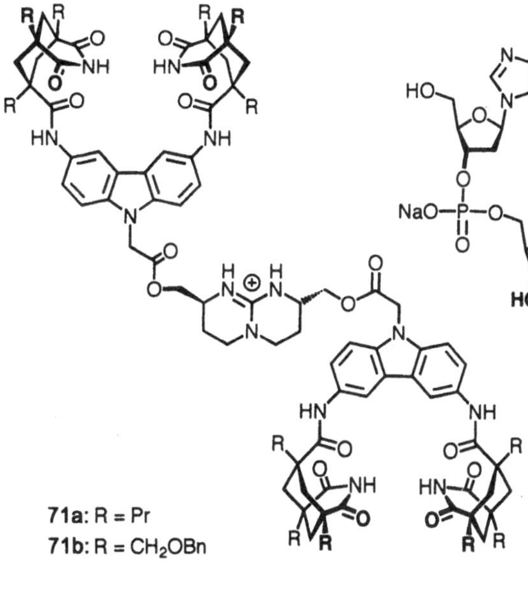

71a: R = Pr
71b: R = CH₂OBn

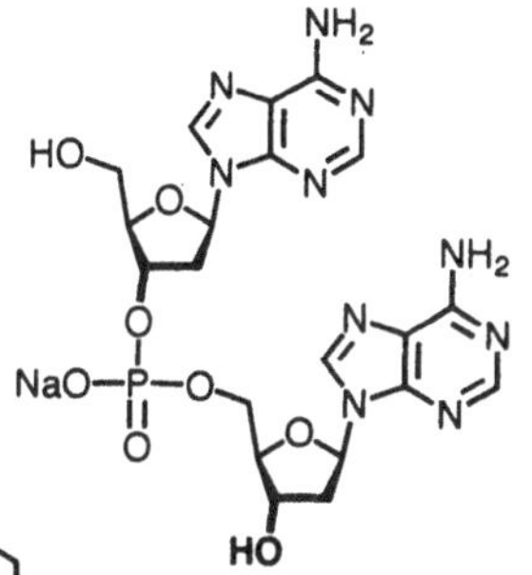

72

The structure of the complex has further been optimized by energy minimizing techniques and subjected to molecular dynamics (Fig. 2) [62]. Nearly the fourth part of the host-guest attraction comes from two strong, nearly coplanar

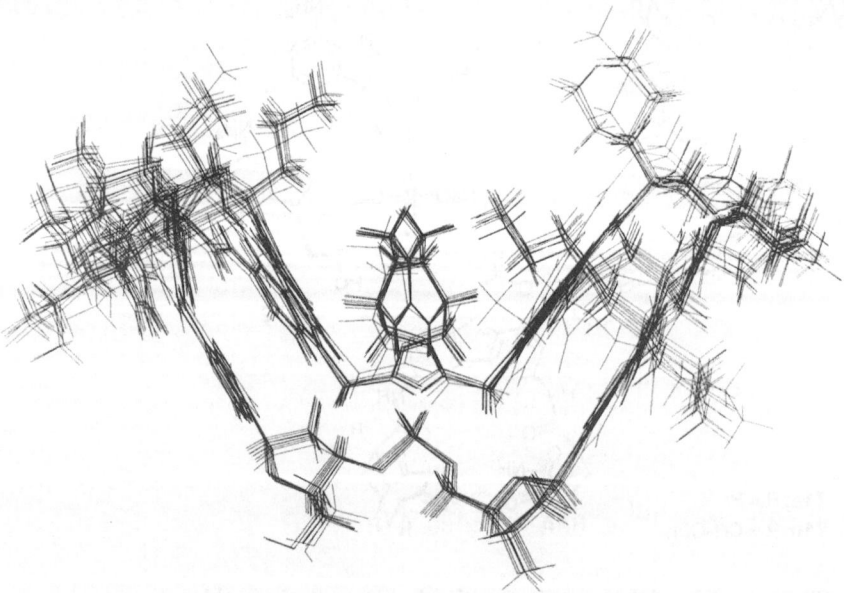

73

hydrogen bonds between the guanidinium and the phosphate groups with average O–H distances of 1.99 and 1.93 Å. The π-stacking between the carbazole and adenine units contributes roughly 30% to the enthalpic interactions. However, the asymmetry of d(ApA) causes slightly different arrangements of the two sides of the complex, with the 3'-end being more favorable (ratio 4:3). This asymmetry also becomes visible in Fig. 2 where the molecular niche complexing the 5'-end of the guest (left side) shows higher conformational fluctuations than the other side.

Fig. 2. Best-fit superposition of nine optimized low-energy structures of the d(ApA) complex **73** taking the dinucleotide atoms as the template

In contrast to the carbazole-purine interactions, the pincer-like hydrogen bonding of the heterobases by the two molecular clefts is almost equally strong on each side and provides further 30% of the binding energy. More precisely it was found in all only six instead of eight possible bridges are formed at a time. Corresponding to the NMR data, in the most stable conformations Hoogsteen-base pairing is favored (NH–N7 distances about 2.1 and 2.3 Å, NH_2–OC distances about 2.2 Å each) and accompanied by a third H-bond on the Watson–Crick side (NH_2–OC distances about 2.2 Å each, but NH–N1 distances 4.0 and 3.6 Å), compare formula **74a**. Full Watson–Crick pairing (formula **74b**) can only transiently be achieved at the expense of the cleavage of one of the hydrogen bonds on the other side. The purine ring is thus flipping between two positions with the equilibrium slightly tending to the Hoogsteen pattern. One explanation for these circumstances is that the spatial distance of the two Kemp units is a little to large for a concerted four-point clamping. A second influence comes from the local charge distribution on the two stacking arenes, that is, the mutual polarization of the π-electron clouds and the resulting dipole-induced dipole interactions that determine the face-to-face arrangement.

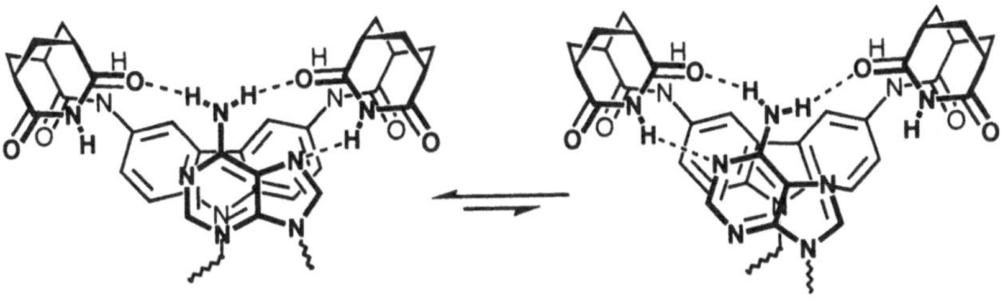

74a   **74b**

The complex of **71a** with d(ApG) has likewise been minimized since this guest also showed affinities in extraction experiments. The overall interaction energy was found to be about $5 \, \text{kcal} \cdot \text{M}^{-1}$ ($21 \, \text{kJ} \cdot \text{M}^{-1}$) inferior compared to the d(ApA) complex. The hydrogen bonded salt bridge is slightly twisted and somewhat less stable. The guanine penetrates into one of the clefts with the six-membered ring only. The NH_2 and N1–H groups act as hydrogen donors and O6 as acceptor. This calculated binding pattern (formula **75**) stands in

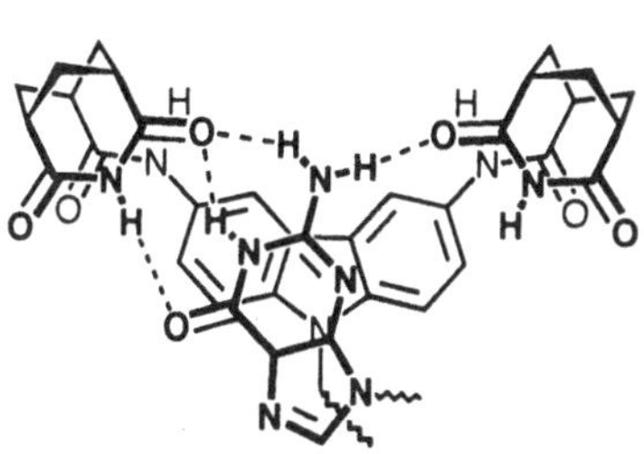

75

contrast to that proposed for the interaction of guanosine and a related molecular niche (without guanidinium unit) [63 p].

More detailed single extraction experiments have been run with host modification **71b**. Not only dinucleotides could be transferred into the organic phase but also a great variety of nucleotides of different chain length, e.g., d(TA)$_3$, d(A$_8$), d(T$_3$G$_2$A$_2$), and even a 76-bases tRNA strand. The only apparent necessity for the guests to be recognized seems to be the presence of at least a few adenines in the strand.

Some selectivity was observed in transport of dinucleotides through a dichloroethylene layer in U-shaped tubes. The highest acceleration rate was determined for d(ApA), whereas d(ApT), d(TpA), and d(ApG) (in this order) were less efficiently carried. For all other examined passengers, even those containing an adenine group, like d(ApC) and d(CpA), no carriage was found.

The huge, multitopic molecule **76**, recently synthesized in our laboratories, was conceived as a receptor for trinucleotides, namely ApApA. The compound consists of three 'adenine pincers' connected by two guanidinium linkers. In a 1:1 complex that is currently under investigation in a further cooperation with the MIT group, ApApA is supposed to be bound by interplay of two electrostatic attraction forces, threefold arene–arene stacking, and thirteen hydrogen bonds in all [104].

With a non-selfcomplementary, zwitterionic host for phosphorylcholine derivatives as the design goal, a bicycloguanidinium building block has been

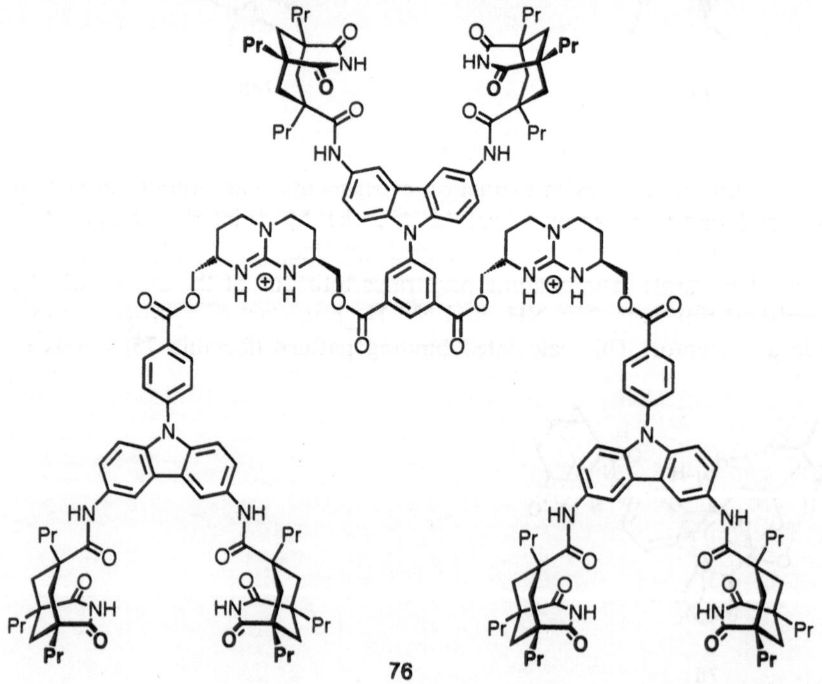

76

attached to a calix[6]arene unit to provide a cavity for the inclusion of the trimethylammonium group. The concept was that in a favorable conformation the side arm of **77** could be located above the opening so that the guest, fixed by a salt bridge, might be attracted into the interior of the cyclophane. Pronounced upfield shifts of the methyl NMR signals in preliminary binding studies with dioctanoylphosphocholine in $CDCl_3$ obviate favorable cation-π interactions between the δ^+-CH_3 groups and the electron-rich framework. Formula **78** outlines the probable structure of the complex [105].

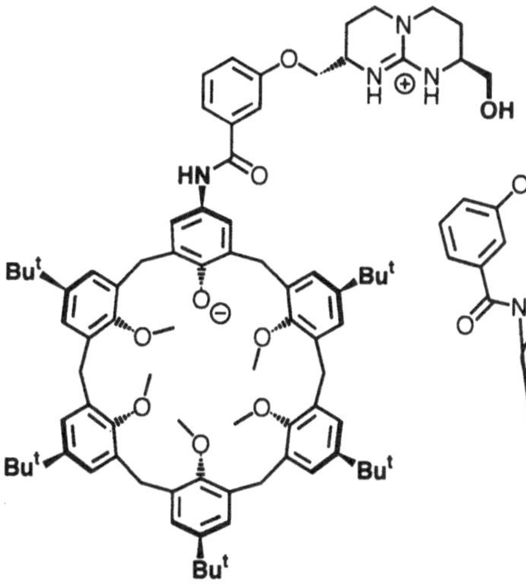

77 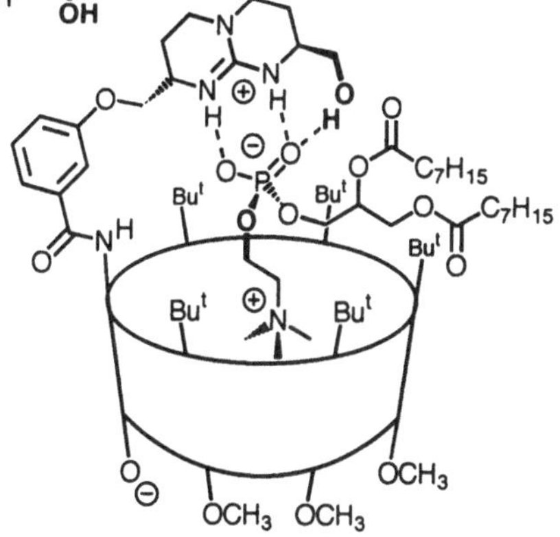   78

5 Concluding Remarks

In 1968, only one year after the publication of the first Pedersen paper on crown ether complexes [106], Park and Simmons marked the corner-stone of anion complexation with the rather fortuitous discovery of halogenide inclusion by macrobicyclic diammonium ions [107]. To label the youngest generation of enzyme mimics and receptor models for amino acids or nucleotides as mere anion ligands, however, would be an over simplification. Nowadays, highly sophisticated host molecules are purposefully synthesized with future applications in mind and their binding behavior is investigated to obtain deeper knowledge about non-covalent intermolecular interactions and to understand better the mechanisms of molecular recognition processes. Supramolecular and bioorgainc chemists increasingly learn how to use hydrogen bonding, hydrophobic effects, π-stacking, and other weak forces as tools for the design of new and advanced compounds that are 'not just molecules' but 'work' and carry out tasks [108]. Nature is often a master and model that ingeniously points the way with its amazingly complicated yet highly perfected molecular systems. The virtuosity with which the melody of life is played in biochemical processes marks a standard still far from being reached artificially by science. With this chapter we intended to summarize the more recent developments in molecular recognition of some types of bio-relevant species and hope to have given the reader an overview of the state of the art in this field.

6 References

1. a) Vögtle F, Sieger H, Müller WM (1981) Top Curr Chem 98: 107; b) Kimura E (1985) Top Curr Chem 128: 113; c) Kimura E (1985) In: Boschke FL (ed) Biomimetic and Biorganic Chemistry. Springer, Berlin Heidelberg New York; d) Schmidtchen FP (1986) Top Curr Chem 132: 101; e) Schmidtchen FP (1988) Nachr Chem Tech Lab 36: 8; f) Dietrich B (1993) Pure Appl Chem 65: 1457
2. Kimura E, Sakonaka A, Yatsunami T, Kodama M (1981) J Am Chem Soc 103: 3041
3. Dietrich B, Hosseini MW, Lehn JM, Sessions RB (1981) J Am Chem Soc 103: 1282
4. Hosseini MW, Lehn JM (1982) J Am Chem Soc 104: 3525
5. The interaction of aliphatic α,ω-dicarboxylates and open-chain benzophenone derivatives containing carboxyl or ammonium groups have been described: Breslow R, Rajagopalan R, Schwarz J (1981) J Am Chem Soc 103: 2905
6. Lehn JM, Méric R, Vigneron JP, Bkouche-Waksman I, Pascard C: J Chem Soc Chem Commun 1991: 62. C.f.: Jazwinski J, Blacker AJ, Lehn JM, Cesario M, Guilhem J, Pascard C (1987) Tetrahedron Lett 28: 6057
7. Dhaenens M, Lehn JM, Vigneron JP: J Chem Soc Perkin Trans 11 1993: 1379
8. Schmidtchen FP (1981) Chem Ber 114: 597; summary: Schmidtchen FP, Gleich A, Schummer A (1989) Pure Appl Chem 61: 1535
9. Schmidtchen FP (1986) J Am Chem Soc 108: 8249
10. Dietrich B, Fyles DL, Fyles TM, Lehn JM (1979) Helv Chim Acta 62: 2763
11. Dietrich B, Fyles TM, Lehn JM, Pease LG, Fyles DL: J Chem Soc Chem Commun 1978: 2763
12. Martell AE, Motekaitis RJ (1988) J Am Chem Soc 110: 8059
13. Tabushi I, Shimizu N, Sugimoto T, Shiozuka M, Yamamura K (1977) J Am Chem Soc 99: 7100
14. Žinić M, Frkanec L, Škarić V, Trafton J, Gokel GW: J Chem Soc Chem Commun 1990: 1726; Žinić M, Frkanec L, Škarić V, Trafton J, Gokel GW (1992) Supramol Chem 1: 47. C.f.: Tsukube H: J Chem Soc Perkin Trans. I 1982: 2359
15. Beer PD, Drew MGB, Hazlewood C, Hesek D, Hodacova J, Stokes SE: J Chem Soc Chem Commun 1993: 229
16. Echavarren A, Galán A, de Mendoza J, Salmerón A, Lehn JM (1988) Helv Chim Acta 71: 685. C.f.: de Mendoza J (1993) An Quim 89: 57
17. Gleich A, Schmidtchen FP (1990) Chem Ber 123: 907; Schmidtchen FP (1990) Tetrahedron Lett 31: 2269; Kurzmeier H, Schmidtchen FP (1990) J Org Chem 55: 3749
18. Schmidtchen FP (1980) Chem Ber 113: 2175
19. Müller G, Riede J, Schmidtchen FP (1988) Angew Chem Int Ed Engl 27: 1516
20. Echavarren A. Galán A, Lehn JM, de Mendoza J (1989) J Am Chem Soc 111: 4994. C.f.: Rebek J Jr (1990) Chemtracts Org Chem 3: 240
21. Galán A, Lu GY, Sánchez J, Seel C, de Mendoza J: unpublished results
22. A review involving enantioselective recognition: Webb TH, Wilcox CS: Chem Soc Rev 1993: 383

23. Schießl P, Schmidtchen FP (1993) Tetrahedron Lett 34: 2449
24. The X-ray crystal structure of the nitrate salt of a parent monoguanidinium bis(silylether) derivative has been published: Gleich A, Schmidtchen FP, Mikulcik P, Müller G: J Chem Soc Chem Commun 1990: 55
25. Chicharro JL, Prados P, de Mendoza J: J Chem Soc Chem Commun 1994: 1193
26. Review: Rebek J Jr (1990) Angew Chem Int Ed Engl 29: 245
27. Rebek J Jr, Nemeth D, Ballester P, Lin FT (1987) J Am Chem Soc 109: 3474
28. Mussons ML, Raposo C, Anaya J, Grande M, Morán JR, Caballero MC: J Chem Soc Perkin Trans 1 1992: 3125
29. Crego M, Raposo C, Caballero MC, García E, Saez JG, Morán JR (1992) Tetrahedron Lett 33: 7437. C.f. Crego M, Raposo C, Partearroyo A, Mussons M, Caballero MC, Morán JR (1993) An Quim 89: 153. C.f. ref. 46g). A related receptor of ureas and formamides is described in: Crego M, Marugán JJ, Raposo C, Sanz MJ, Alcázar, V, Caballero MC, Morán JR (1991) Tetrahedron Lett 32: 4185
30. Reviews: a) Hamilton AD (1991) In: Dugas H (ed) Bioorganic Chemistry Frontiers, vol 2. Springer, Berlin Heidelberg New York, p 115; b) Hamilton AD, Fan E, Van Arman S, Vicent C, Garcia Tellado F, Geib SJ (1993) Supramol Chem 1: 247
31. Garcia Tellado F, Goswami S, Chang SK, Geib SJ, Hamilton AD (1990) J Am Chem Soc 112: 7393; therein is also described the complex of a macromonocycle with diethylmalonate. Carboxylic acid complexation by a synthetic vancomycin analogue: Pant N, Hamilton AD (1988) J Am Chem Soc 110: 2002
32. Vicent C, Hirst SC, Garcia Tellado F, Hamilton AD (1991) J Am Chem Soc 113: 5466
33. Related mixed urea-amidopyridyl or tetraamidic hosts have been reported by the same group: Vicent C, Fan E, Hamilton AD (1992) Tetrahedron Lett 33: 4269; Albert JS, Hamilton AD (1993) Tetrahedron Lett 34: 7363. For a bis(urea) host compound, see: Hamann BC, Branda NR, Rebek J Jr (1993) Tetrahedron Lett 34: 6837
34. Fan E, Van Arman SA, Kincaid S, Hamilton AD (1993) J Am Chem Soc 115: 369
35. Some recent papers on this topic: a) Lindsey JS (1991) New J Chem 15: 153; b) Simard M, Su D, Wuest JD (1991) J Am Chem Soc 113: 4696; c) Whitesides GM, Mathias JP, Seto CT (1992) Science 254: 1312; d) Duerr BF, Zimmermann SC (1992) J Org Chem 57: 2215; e) Bonar-Law RP, Sanders JKM (1993) Tetrahedron Lett 34: 1677; f) Seto CT, Whitesides GM (1993) J Am Chem Soc 115: 1330 and references cited therein; g) Potts KT, Gheysen Raiford KA, Keshavarz-K M (1993) J Am Chem Soc 115: 2793; h) Chang YL, West MA, Fowler FW, Lauher JW (1993) J Am Chem Soc 115: 5991; i) Piguet C, Bünzli JCG, Bernardinelli G, Hopfgartner G, Williams AF (1993) J Am Chem Soc 115: 8197; j) Baxter P, Lehn JM, DeCian A, Fischer J (1993) Angew Chem Int Ed Engl 32: 69; k) Krämer R, Lehn JM, DeCian A, Fischer J (1993) Angew Chem Int Ed Engl 32: 703; l) Ziessel R, Youinou MT (1993) Angew Chem Int Ed Engl 32: 877; m) Armspach D, Ashton PR, Moore CP, Spencer N, Stoddart JF, Wear TJ, Williams DJ (1993) Angew Chem Int Ed Engl 32: 854; n) Rissanen K, Huuskonen J, Windscheif PM, Vögtle (1993) Supramol Chem 2: 247; o) Wyler R, de Mendoza J, Rebek J Jr (1993) Angew Chem Int Ed Engl 32: 1699; p) Andreu C, Beerli R, Branda N, Conn M, de Mendoza J, Galán A, Huc I, Tymoshenko M, Valdez C, Winter E, Wyler R, Rebek J Jr (1993) Pure Appl Chem 65: 2313; q) Mathias JP, Simanek EE, Seto CT, Whitesides GM (1993) Angew Chem Int Ed Engl 32: 1766; r) Ghadiri MR, Granja JR, Milligan RA, McRee DE, Khazanovich (1993) Nature 366: 324
36. Garcia Tellado F, Geib SJ, Goswami S, Hamilton AD (1991) J Am Chem Soc 113: 9265
37. Geib SJ, Vicent C, Fan E, Hamilton AD (1993) Angew Chem Int Ed Engl 32: 60
38. Yang J, Fan E, Geib SJ, Hamilton AD (1993) J Am Chem Soc 115: 5314
39. Alcázar Montero V, Tomlinson L, Houk KN, Diederich F (1991) Tetrahedron Lett 32: 5309; Alcázar V, Morán JR, Diederich F (1992) Israel J Chem 32: 69
40. Owens L, Thilgen C, Diederich F, Knobler CB (1993) Helv Chim Acta 76: 2757
41. Alcázar V, Diederich F (1992) Angew Chem Int Ed Engl 31: 1503
42. Garcia Tellado F, Albert J, Hamilton AD: J Chem Soc Chem Commun 1991: 1761
43. Examples: a) Zielinsky WS, Orgel LE (1987) Nature 327: 346; b) von Kiedrowski G, Wlotzka

C. Seel et al.

B, Helbing J, Matzen M, Jordan S (1991) Angew Chem Int Ed Engl 30: 423 and references herein; c) Nowick JS, Feng Q, Tjivikua T, Ballester P, Rebek J Jr (1991) J Am Chem Soc 113: 8831; a summary is given in: Rebek J Jr (1993) Supramol Chem 1: 261 d) Krämer R, Lehn JM, De Cian A, Fischer J (1993) Angew Chem Int Ed Engl 32: 764

44. Terfort A, von Kiedrowski G (1992) Angew Chem Int Engl 31: 654
45. a) Pederson noted that 18-crown-6 binds glycine hydrochloride but not glycine itself: Pederson CJ (1967) J Am Chem Soc 89: 7017; b) Lehn reports similar observations with phenylalanine: Behr JP, Lehn JM, Vierling P (1982) Helv Chim Acta 65: 1853
46. a) Molecular clefts: Famulok M, Jeong KS, Deslongchamps G, Rebek J Jr (1991) Angew Chem Int Ed Engl 30: 858; b) Macrooligocyclic hosts: Liu R, Sanderson PEJ, Still WC (1990) J Org Chem 55: 5184; Yoon SS, Still WC (1993) J Am Chem Soc 115: 823; Liu R, Still WC (1993) Tetrahedron Lett 34: 2573; Yoon SS, Georgiadis TM, Still WC (1993) Tetrahedron Lett 34: 6697; Borchardt A, Still WC (1994) J Am Chem Soc 116: 373; c) Short overview: Schneider HJ (1993) Angew Chem Int Ed Engl 32: 848; d) Zinc porphyrins: Kuroda Y, Kato Y, Higashioji T, Ogoshi H (1993) Angew Chem Int Ed Engl 32: 723; e) Calix[6]arenes: Chang SK, Hwang HS, Son H, Youk J, Kang YS: J Chem Soc Chem Commun 1991: 217; f) Macrocyclic peptides: Miyake H, Kojima Y, Yamashita T, Ohsuka A (1993) Macrocyclic Chem 194: 1925; g) Chiral molecular niches related to host 21: Crego M, Partearroyo A, Raposo C, Müssons ML López JL, Alcázar V, Morán JR (1994) Tetrahedron Lett 35: 1435
47. Peacok SS, Walba DM, Gaeta FCA, Helgeson RC, Cram DJ (1980) J Am Chem Soc 104: 2043
48. Goodnow TT, Reddington MV, Stoddart JF, Kaifer AE (1991) J Am Chem Soc 113: 4335
49. Kobayashi K, Tominaga M, Asakawa Y, Aoyama Y (1993) Tetrahedron Lett 34: 5121. For chiral recognition of N-protected amino acids by a metalloporphyrin, see: Konishi K, Yahara K, Toshishige H, Aida T, Inoue S (1994) J Am Chem Soc 116: 1337
50. Tabushi I, Kuroda Y, Mizutani T (1986) J Am Chem Soc 108: 4514. It should be noticed that unsubstituted α- and β-cyclodextrin as well weakly bind tryptophan: see therein and Lipkowitz KB, Raghothama S, Yang YA (1992) J Am Chem Soc 114: 1554
51. Impellizzeri G, Maccarrone G, Rizzarelli E, Vecchio G, Corradini R, Marchelli R (1991) Angew Chem Int Ed Engl 30: 1348. The capacity of the aminoacetate subunit of α-amino acids to chelate Lewis-acidic centers has also been used in U-tube experiments with lipophilic Cu(II) complexes (a) and phenylboronic acid (b), respectively, as transport mediators: a) Scrimin P, Tonellato U, Zanta N (1988) Tetrahedron Lett 29: 4967; b) Mohler LK, Czarnik AW (1993) J Am Chem Soc 115: 7037
52. Aoyama Y, Asakawa M, Yamagishi A, Toi H, Ogoshi H (1990) J Am Chem Soc 112: 3145
53. Sunamoto J, Iwamoto K, Mohri Y, Kominato T (1982) J Am Chem Soc 104: 5502
54. Rebek J Jr, Nemeth D (1985) J Am Chem Soc 107: 6738; Rebek J Jr, Askew B, Nemeth D, Parris K (1987) J Am Chem Soc 109: 2432
55. Amino acid binding by monolayers of long-chain derivatives of Kemp's triacid: Ikeura Y, Kurihara K, Kunitake T (1991) J Am Chem Soc 113: 7342
56. Askew BC (1990) Tetrahedron Lett 30: 4245
57. Murakami Y, Ohno T, Hayashida O, Hisaeda Y: J Chem Soc Chem Commun 1991: 950. C.f. Murakami Y, Hayashida O, Ito T, Hisaeda Y: Chem Lett 1992: 497. Murakami Y, Hayashida O Nagai Y (1994) J Am Chem Soc 116: 2611
58. A similar open-chain host binding hydrophobic anionic and neutral dyes in water: Murakami Y, Hayashida O, Nagai Y, (1993) Tetrahedron Lett 34: 7935
59. Schmidtchen FP (1986) J Org Chem 51: 5161
60. Galán A, Andreu D, Echavarren AM, Prados P, de Mendoza J (1992) J Am Chem Soc 114: 1511
61. More recent HPLC measurements with eluents containing chiral salts showed that ca. 10% of the minor enantiomers were actually extracted: Marcelli R, de Mendoza J, et al. (unpublished results)
62. de Mendoza J, Gago F (1994) In: Wipff G (ed) Computational Aproaches in Supramolecular Chemistry. Kluwer, Dordrecht, p 79

63. a) Hamilton AD, Van Engen D (1987) J Am Chem Soc 109: 5035; b) Kelly TR, Maguire MP (1987) J Am Chem Soc 109: 6549; c) Feibush B, Saha M, Onan K, Karger B, Giese R (1987) J Am Chem Soc 109: 7531; d) Muehldorf AV, Van Engen D, Warner JC, Hamilton AD (1988) J Am Chem Soc 110: 6561; e) Lindsey JS, Kearney PC, Duff RJ, Tjivuaka T, Rebek J Jr (1988) J Am Chem Soc 110: 6575; f) Grotjohn BF, Czarnik AW (1989) Tetrahedron Lett 30: 2325; g) Goswami S, Hamilton AD, Van Engen D (1989) J Am Chem Soc 111: 3425; h) Zimmerman SC, Wu W (1989) J Am Chem Soc 111: 8054; i) Adrian JC, Wilcox CS (1989) J Am Chem Soc 111: 8055; j) Jeong KS, Tjivikua T, Muehldorf A, Deslongchamps G, Famulok M, Rebek J Jr (1991) J Am Chem Soc 113: 201; k) Medina JC, Li C, Bott SG, Atwood JL, Gokel GW (1991) J Am Chem Soc 113: 366; l) Seel C, Vögtle F (1991) Angew Chem Int Ed Engl 30: 442; m)Park TK, Schroeder J, Rebek J Jr (1991) J Am Chem Soc 113: 5125; n) Ogoshi H, Hatekeyama H, Kotani J, Kawashima A, Kuroda Y (1991) J Am Chem Soc 113: 8181; o) Murray TJ, Zimmerman SC (1992) J Am Chem Soc 114: 4010; p) Conn MM, Deslongchamps G, de Mendoza J, Rebek J Jr (1993) J Am Chem Soc 115: 3548

64. Some recent papers involving phosphate ligands: a) Huston ME, Akkaya EU, Czarnik AW (1989) J Am Chem Soc 111: 8735; b) Rudkevich DM, Stauthamer WPRV, Verboom W, Engbersen JFJ, Harkema S, Reinhoudt DN (1992) J Am Chem Soc 114: 9671; c) Sessler JL, Cyr M, Furuta H, Král V, Mody T, Morishima T, Shonoya M, Weghorn S (1993) Pure Appl Chem 65: 393; d) Valiyaveettil S, Engbersen JFJ, Verboom W, Reinhoudt DN (1993) Angew Chem Int Ed Engl 32: 900; e) Rudkevich DM, Brzozka Z, Palys M, Visser HC, Verboom W, Reinhoudt DN (1994) Angew Chem Int Ed Engl 33: 467

65. a) Nakai C, Glinsman W (1977) Biochemistry 16: 5636; b) Dietrich B, Hosseini MW, Lehn JM, Sessions RB (1981) Biochemistry 103: 1282; c) Kimura E, Kodama M, Yatsunami T (1982) J Am Chem Soc 104: 3182; d) Marecek JF, Fischer PA, Burrows CJ (188) Tetrahedron Lett 29: 6231

66. a) Behr JP: J Chem Soc Chem Commun 1989: 101 and references cited therein; b) Zuber G, Sirlin C, Behr JP (1993) J Am Chem Soc 115: 4939; c) Van Arman SA, Czarnik AW (1990) J Am Chem Soc 112: 5376; d) Wilson HR, Williams RJP (1987) J Chem Soc Faraday Trans 1 83: 1885

67. Hosseini MW, Lehn JM, Jones KC, Plute KE, Mertes KB, Mertes MP (1989) J Am Chem Soc 111: 6330 and references cited therein

68. Hosseini MW, Blacker AJ, Lehn JM: J Chem Soc Chem Commun 1988: 596; Hosseini MW, Blacker AJ, Lehn JM (1990) J Am Chem Soc 112: 3896. C.f. Mertes MP, Mertes KB (1990) Acc Chem Res 23: 413

69. Fluorescence monitoring of enzymatic ATP hydrolysis by an anthrylpolyammonium ion is described in: Van Arman SA, Czarnik AW (1993) Supramol Chem 1: 99

70. For recently published binding of nucleotides by aminocyclodextrins and a water soluble rhodioporphyrin, respectively, see: a) Eliseev AV, Schneider HJ (1993) Angew Chem Int Ed Engl 32: 1331; b) Kuroda Y, Hatakeyama H, Inakoshi N, Ogoshi H (1993) Tetrahedron Lett 34: 8285

71. Claude S, Lehn JM, Schmidt F, Vigneron JP: J Chem Soc Chem Commun 1991: 1182

72. Schneider HJ, Blatter T, Palm B, Pfingstag U, Rüdiger V, Theis I (1992) J Am Chem Soc 114: 7704

73. a) Harden MR (ed) (1985) Approaches to antiviral agents. VCH Publishers, Deerfield Beach, Florida; b) Martin JC (ed) (1989) Nucleotide analogues as antiviral agents. Am Chem Soc, Washington, D.C.

74. Mitsuya H, Broder S (1987) Nature 325: 773

75. Tabushi I, Imuta JI, Seko N, Kobuke Y (1978) J Am Chem Soc 100: 6287; Tabushi I, Kobuke Y, Imuta JI (1980) J Am Chem Soc 102: 1744: Tabushi I, Kobuke Y, Imuta JI (1981) J Am Chem Soc 103: 6152

76. Li T, Diederich F (1992) J Org Chem 57: 3449

77. Li T, Krasne SJ, Persson B, Kaback HR, Diederich F (1993) J Org Chem 58: 380

78. Furuta H, Magda D, Sessler JL (1991) J Am Chem Soc 113: 978

C. Seel et al.

79. Furuta H, Cyr MJ, Sessler JL (1991) J Am Chem Soc 113: 6677. C.f. Iverson BL, Thomas RE, Kral V, Sessler JL (1994) J Am Chem Soc 116: 2663
80. Král V, Sessler Jl, Furuta H (1992) J Am Chem Soc 114: 8704
81. A review summarizing this work and a detailed bibliography is given in: Sessler JL, Furuta H, Král V (1993) Supramol Chem 1: 209
82. Furuta H, Morishima T, Král V, Sessler JL (1992) Supramol Chem 3: 5
83. Manabe K, Okamura K, Date T, Koga K (1992) J Am Chem Soc 114: 6940
84. Flatt LS, Lynch V, Anslyn EV (1992) Tetrahedron Lett 33: 2785
85. Geib SJ, Hirst SC, Vicent C, Hamilton AD: J Chem Soc Chem Commun 1991: 1283
86. Dixon RP, Geib SJ, Hamilton AD (1992) J Am Chem Soc 114: 365
87. Tecilla P, Chang SK, Hamilton AD (1990) J Am Chem Soc 112: 9586
88. Jubian V, Dixon RP, Hamilton AD (1992) J Am Chem Soc 114: 1120
89. Cotton FA, Hazen EE, Legg MJ (1979) Proc Natl Acad Sci USA 64: 420
90. Ariga K, Anslyn EV (1992) J Org Chem 57: 417. For the enolate complexation of a related polyaza-cleft, see: Kelly-Rowley AM, Cabell LA, Anslyn EV (1991) J Am Chem Soc 113: 9687
91. For the proposed hydrolysis mechanism of the enzyme, see: a) Sepersu EH, Shortle D, Mildvan AS (1987) Biochemistry 26: 1289; b) Aqvist J, Warshel A (1990) J Am Chem Soc 112: 2860
92. Smith J, Ariga K, Anslyn EV (1993) J Am Chem Soc 113: 362; Kneeland DM, Ariga K, Lynch VM, Huang CY, Anslyn EV (1993) J Am Chem Soc 115: 10042
93. This work is summarized in: Anslyn EV, Smith J, Kneeland DM, Ariga K, Chu FY (1993) Supramol Chem 1: 201
94. Breslow R, Doherty JB, Guillot G, Lipsey Hersh C (1978) J Am Chem Soc 100: 3227; Breslow R, Bovy P, Lipsey Hersh C (1980) J Am Chem Soc 102: 2115; Breslow R, Labelle M (1986) J Am Chem Soc 108: 2655; Anslyn EV, Breslow R (1989) J Am Chem Soc 111: 4473; Anslyn EV, Breslow R (1989) J Am Chem Soc 111: 5972. Summary: Breslow R (1993) Supramol Chem 1: 111
95. The literature concerning catalytic phosphoester and polynucleotide hydrolysis is extended. For recent publications with macrocyclic complexes as agents and a detailed bibliography, see: a) Morrow JR, Buttrey LA, Shelton VM, Berback KA (1992) J Am Chem Soc 114: 1903; b) Göbel MW, Bats JW, Dürner G (1992) Angew Chem Int Ed Engl 31: 207; c) Vance DH, Czarnik AW, (1993) J Am Chem Soc 115: 12165; d) Schneider HJ, Rammo J, Hettich R (1993) Angew Chem Int Ed Engl 32: 1716
96. ATP binding by a guanidinium-functionalized monolayer: Sasaki DY, Kurihara K, Kunitake T (1991) J Am Chem Soc 113: 9685
97. Galán A (1989) Thesis, Universidad Autónoma de Madrid, Madrid
98. Schmidtchen FP (1989) Tetrahedron Lett 30: 4493
99. Schießl P, Schmidtchen FP (1994) J Org Chem 59: 509
100. Galán A, Pueyo E, Salmerón A, de Mendoza J (1991) Tetrahedron Lett 32: 1827
101. Deslongchamps G, Galán A, de Mendoza J, Rebek J Jr (1992) Angew Chem Int Ed Engl 31: 61
102. Andreu C, Galán A, Kobiro K, de Mendoza J, Park TK, Rebek J Jr, Salmerón A, Ussman N (1994) J Am Chem Soc 166: 5501
103. Galán A, de Mendoza J, Toiron C, Bruix M, Deslongchamps G, Rebek J Jr (1991) J Am Chem Soc 113: 9424
104. Salmerón A, de Mendoza J, Toiron C, Bruix M, Conn MM and Rebek J, Jr.: unpublished results
105. Magrans JO, de Mendoza J, Prados P, Sánchez J, Seel C (unpublished results)
106. Pedersen CJ (1967) J Am Chem Soc 89: 2495
107. Park CH, Simmons HE (1968) J Am Chem Soc 90: 2431
108. "The primary motivations that once induced chemists to undertake natural product syntheses no longer exist. Instead of target structures themselves, molecular function and activity now occupy center stage." Quoted from: Seebach D (1990) Angew Chem Int Ed Engl 29: 1320

Molecular Recognition by Large Hydrophobic Cavities Embedded in Synthetic Bilayer Membranes

Yukito Murakami,[1,2] Jun-ichi Kikuchi,[2] and Osamu Hayashida[1]

[1] Department of Chemical Science and Technology, Faculty of Engineering, Kyushu University, Fukuoka 812, Japan
[2] Institute for Fundamental Research in Organic Chemistry, Kyushu University, Fukuoka 812, Japan

Table of Contents

Topics in Current Chemistry, Vol. 175
© Springer-Verlag Berlin Heidelberg 1995

Artificial receptors of three different structural modes, each being capable of providing a large hydrophobic cavity, were designed and synthesized: octopus cyclophanes having eight flexible hydrocarbon branches, steroid cyclophanes bearing four rigid steroid moieties, and cage-type cyclophanes composed of two macrocyclic skeletons and four chiral bridges connecting both macrocycles. In aqueous solution these cationic cyclophanes are effective in molecular recognition toward anionic and nonionic organic guests as artificial intracellular receptors. Hybrid assemblies are formed by combination of the individual cyclophanes with synthetic bilayer membranes composed of a peptide lipid that covalently involves an L-alanine residue interposed between an anionic head group and a hydrophobic double-chain segment. Such hybrid assemblies are considered to be artificial cell-surface receptors and generated supramolecular effects on molecular recognition.

1 Introduction

Biological cells are the most ingeniously designed supramolecular devices in nature. Many metabolic processes in cells are considered to be triggered through recognition of signal molecules by specific proteins, so-called receptors. The receptor concept born in the early part of this century [1, 2] has become necessary to be characterized in view of the rapid growth of molecular biology [3]. Biological receptors are classified into two categories depending on their location in cells: intracellular receptors in cytoplasms and cell-surface receptors in biomembranes. Furthermore, there are at least three known classes of cell-surface receptors – channel-linked, G-protein-linked, and catalytic ones. However, receptor functions remain to be clarified at the molecular level with emphasis on molecular recognition and the resulting responses.

Studies on molecular recognition by artificial receptors are thus one of the most important approaches to such characterization in relation to supramolecular chemistry [4]. Functional simulation of intracellular receptors in aqueous media has been actively carried out with attention to various noncovalent host–guest interactions, such as hydrophobic, electrostatic, hydrogen-bonding, charge-transfer, and van der Waals modes [5–10]. On the other hand, molecular recognition by artificial cell-surface receptors embedded in supramolecular assemblies has been scarcely studied up to the present time, except for channel-linked receptors [11–13].

We have recently developed macrocyclic compounds as artificial receptors, each having a three-dimensionally extended hydrophobic cavity [14]. In aqueous media, octopus cyclophanes bearing eight hydrocarbon branches introduced into a cyclophane skeleton provide large and flexible hydrophobic cavities capable of exhibiting guest recognition through the induced-fit mechanism

[15–17]. On the other hand, the hydrophobic internal cavity provided by a cage-type cyclophane, Kyuphane, is suitable for molecular recognition through the biomimetic lock-and-key mechanism [18–22]. We will discuss specific molecular recognition behavior of novel artificial receptors of three different structural modes: octopus, steroid, and cage-type cyclophanes. First, we describe molecular design and syntheses of these cyclophane derivatives, and their molecular recognition ability as artificial intracellular receptors in aqueous media.

We have already reported that synthetic peptide lipids, having α-amino acid residue(s) interposed between a polar head moiety and a hydrophobic double-chain segment, can be used as models for functional simulation of biomembranes [23]. On this ground, we are to clarify molecular recognition specificity by supramolecular assemblies formed in combination of the macrocyclic receptors with the peptide lipid as artificial cell-surface receptors.

2 Design and Syntheses of Supramolecular Elements

Steroid hormones are bound reversibly to specific receptor proteins, typical intracellular receptors in cytoplasms and nuclei. The binding of hormone activates the receptor, enabling it to bind with high affinity to specific DNA sequences that act as transcriptional enhancers. On the other hand, receptor proteins placed in cell surfaces bind various signal molecules, such as neuro-transmitters, protein hormones, and growth factors, with high affinity and convert these extracellular events into intracellular signals that alter the behavior of the target cell [3]. Thus, recognition of signal molecules by receptors is a common and basic phenomenon in the cell signaling. We designed macrocyclic hosts which are capable of simulating guest-binding behavior not

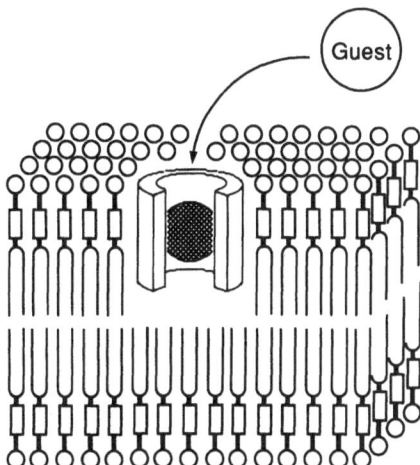

Fig. 1. Schematic representation of molecular recognition by an artificial receptor embedded in a bilayer membrane

only of artificial intracellular receptors but also of cell-surface ones (Fig. 1). In addition, design and syntheses of peptide lipids, which are capable of forming hybrid assemblies with these artificial receptors, are to be mentioned briefly.

2.1 Artificial Receptors

The overall guest-binding ability of a host molecule in aqueous media is highly dependent on the hydrophobic character of its cavity, and is enhanced as the hydrophobicity increases since noncovalent host–guest interactions become more effective in well-desolvated and hydrophobic microenvironments. Although moderate guest recognition has been performed by various cyclophanes composed of a single macrocyclic skeleton, more efficient molecular recognition can be achieved by modified cyclophane hosts capable of providing a three-dimensionally extended internal cavity [14]. Moreover, it is required that these hosts are sufficiently hydrophobic so as to form hybrid assemblies with bilayer-

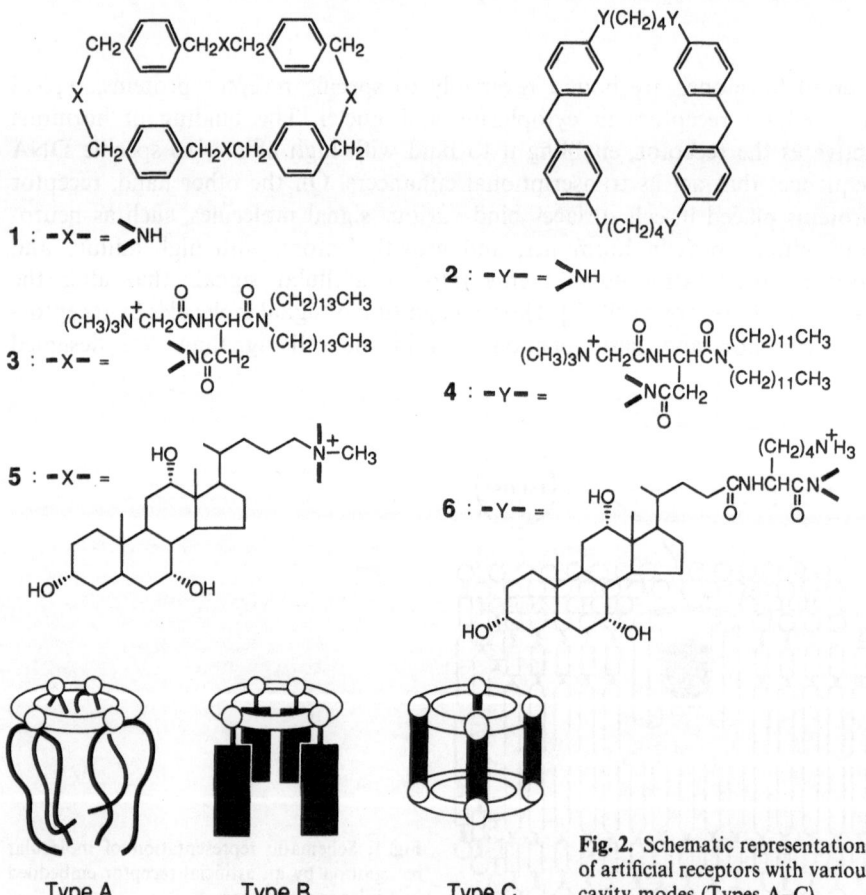

Fig. 2. Schematic representations of artificial receptors with various cavity modes (Types A–C)

forming lipids. On these grounds, we performed three-types of modifications on tetraaza[3.3.3.3]paracyclophane (1) and tetraaza[6.1.6.1]paracyclophane (2), as schematically shown in Fig. 2: type **A**, azacyclophanes having eight flexible hydrocarbon chains (octopus cyclophanes); type **B**, azacyclophanes bearing four rigid steroid moieties (steroid cyclophanes); type **C**, linked azacyclophanes in which two azacyclophane skeletons are connected with four chiral and hydrophobic bridging components (cage-type cyclophanes).

2.1.1 Octopus Cyclophanes

Both of the tetraaza[3.3.3.3]paracyclophane (1) and tetraaza[n.1.n.1]paracyclophane (n = 6, 7, 8; cf. **2**) rings have frequently been used as fundamental molecular skeletons for preparation of functionalized macrocyclic hosts [24–36]. Formation of three-dimensionally extended hydrophobic cavities was approached by introducing multiple hydrocarbon branches into the macrocyclic skeletons. Multiple hydrophobic chains thus placed in a macrocycle must be extended in the same direction and undergo mutual association to attain their optimal hydrophobic interactions in aqueous media due to thermodynamic reasons, while in nonaqueous media they presumably assume a free and separated configuration to minimize their mutual steric interactions. Consequently, such hydrophobic branches may provide a large hydrophobic cavity in aqueous media.

First, we prepared so-called octopus-like cyclophanes having four alkyl branches introduced into the tetraaza[3.3.3.3]paracyclophane ring [25, 26]. These cyclophanes can incorporate hydrophobic guests in aqueous media, exhibiting the induced-fit binding behavior. In order to enhance the guest-binding ability, we have prepared real octopus cyclophanes having eight hydrophobic chains [15, 16]. These octopus cyclophanes covalently involve L-lysine residues as connector units interposed between a rigid tetraaza[3.3.3.3]paracyclophane skeleton and four double-chain hydrocarbon segments. Recently, we designed and synthesized novel octopus cyclophanes composed of a macrocyclic tetraaza[3.3.3.3]- or a tetraaza[6.1.6.1]paracyclophane ring, four double alkyl-chains, and L-aspartate residues interposed between them (**3** and **4**) [17, 37]. The space-filling models of these octopus cyclophanes, as generated by the BIOGRAF software, strongly suggest that each host molecule incorporates hydrophobic guests of various bulkiness into its hydrophobic cavity that is well shielded from the bulk aqueous phase (Fig. 3A).

2.1.2 Steroid Cyclophanes

Steroids of biological importance, such as sex hormones and adrenal cortical hormones, cardiac glycosides, and bile salts, are widely distributed in living cells. Among these steroids, the bile salts act as hosts which play crucial roles in digestion of fats and in pathogenesis of cholesterol gallstones. Such physiological activity originates from their ability to form micelles and small molecular

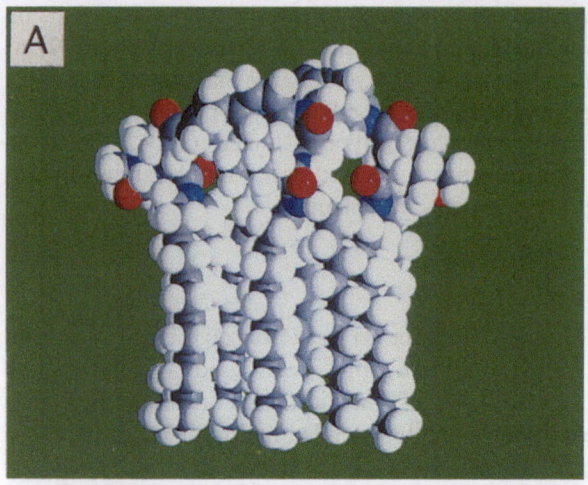

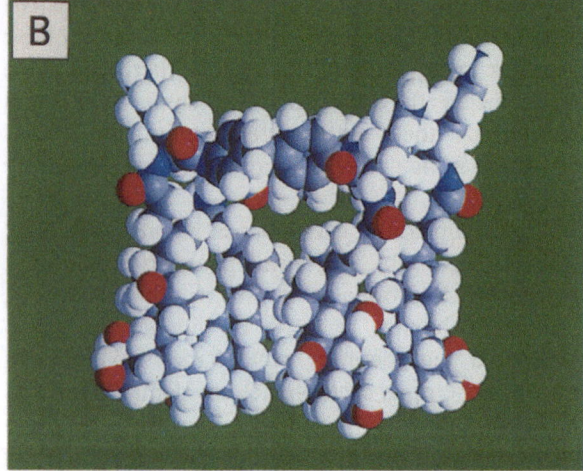

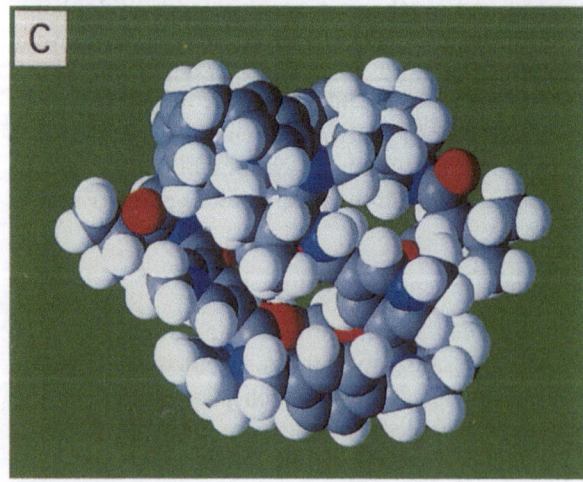

Fig. 3A–C. Space-filling models of artificial receptors, each being capable of providing a large hydrophobic cavity: **A** octopus cyclophane **4**; **B** steroid cyclophane **6**; **C** cage-type cyclophane **8**

aggregates which can solubilize hydrophobic species, such as fatty acids, monoglycerides, phospholipids, and cholesterol. We have designed cyclophane derivatives bearing four cholate moieties (**5** and **6**), so-called steroid cyclophanes, which can be regarded as functional models for the bile salt micelle [37–39]. The computer-aided molecular model study reveals that each of the steroid cyclophanes is able to incorporate a hydrophobic guest molecule into its three-dimensionally extended inner cavity created by the four steroid moieties and the macrocyclic skeleton in a manner similar to that performed by the octopus cyclophanes (Fig. 3B). It needs to be noted, however, that the specific structural rigidity of the steroid moiety in the former host, as compared with the flexible nature of the hydrocarbon chain in the latter, may be reflected in the characteristics of molecular recognition ability.

2.1.3 Cage-Type Cyclophanes

In order to construct a hydrophobic three-dimensional cavity that is intramolecularly limited in space, we have prepared cage-type cyclophanes by linking macrocyclic rings. First we prepared a macropolycyclic host, which is constructed with two rigid macrocyclic skeletons of different size, tetraaza[3.3.3.3]paracyclophane as the larger one and tetraazacyclotetradecane as the smaller one, and four flexible hydrocarbon chains that connect the two macrocycles [40]. The flexibility of four hydrocarbon chains connecting the two macrocycles allows the induced-fit host–guest interaction in aqueous media.

We have also prepared a cage-type macrocycle, so-called 'Kyuphane', which provides a relatively rigid and large hydrophobic cavity surrounded by six faces, each consisting of a tetraaza[3.3.3.3]paracyclophane ring [18–22]. A large internal cavity of Kyuphane was confirmed to be suitable for molecular recognition through the biomimetic lock-and-key mechanism.

Meanwhile, binding sites of naturally occurring receptors are constructed with various optically active amino acid residues so that those host molecules demonstrate outstanding chiral recognition toward external substances. Under such circumstances, the next strategy is to get further insights into the chirality-based molecular recognition behavior of cage-type cyclophanes toward various guests. We designed a novel cage-type peptide cyclophane bearing chiral binding sites provided by L-valine residues (**7**) [41–47]. The present host molecule is constructed with two rigid macrocyclic skeletons, tetraaza[6.1.6.1]paracyclophane as the larger ring and tetraaza[3.3.3.3]paracyclophane as the smaller one, and four bridging components that connect these macrocycles. Each of the bridging components is composed of a pyridine-3,5-dicarbonyl moiety and a L-valine residue. Host **7** was synthesized by condensation of N,N',N'',N'''-tetrakis(5-carboxynicotinoyl)-2,11,20,29-tetraaza[3.3.3.3]paracyclophane with a peptide cyclophane bearing four L-valine moieties, N,N',N'',N'''-tetrakis(2-aminoisovaleryl)-1,6,20,25-tetraaza[6.1.6.1]paracyclophane, in the presence of diethyl cyanophosphonate (DECP) under high dilution conditions at 0 °C. The product is soluble in acidic aqueous media and behaves as a polycationic host. In

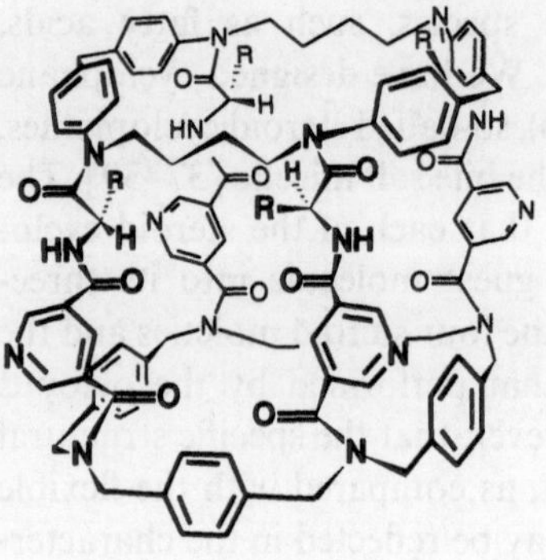

R = CH(CH₃)₂

7

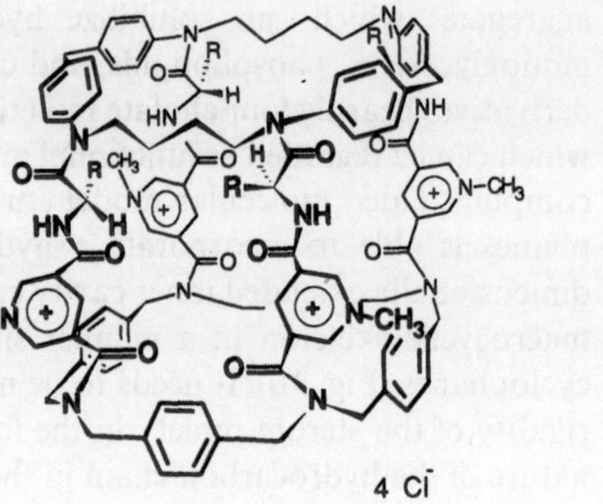

R = CH(CH₃)₂

8

order to obtain water-soluble hosts in aqueous media over a wide pH range, cage-type peptide cyclophane **8** was derived from **7** by quaternization with methyl iodide, followed by replacement of the counterion iodide with chloride [42, 44].

Lowest energy conformations of hosts **7** and **8** in the gas phase were searched by means of molecular mechanics and dynamics calculations (BIOGRAF, Dreiding-I and Dreiding-II [48]). The result reveals that host **8** provides a globular hydrophobic cavity with a maximum inner diameter of ca. 1.0 nm (Fig. 3C). In addition, the four pyridinium moieties bound to the chiral valine residues in the bridging components approach close to each other and are twisted in the same direction. Such a helical conformation for **8** seems to originate from the chiral nature of the valine residue in the bridging segments. A similar asymmetric character of the internal cavity of host **7** was also confirmed by the identical computational method.

2.2 Bilayer-Forming Lipids

It has been proposed that a biomembrane assumes a tripartite structure: a hydrophobic domain composed of aliphatic double-chains, a polar domain composed of hydrophilic head groups, and a hydrogen-belt domain interlaid between these two [49]. Recently, we have developed synthetic lipids forming bilayer membranes with significant morphological stability, so-called synthetic peptide lipids, on the basis of the tripartite concept [23]. Lipids **9** and **10** are the typical anionic and cationic peptide lipids, respectively. The α-carboxylato and α-amino groups of an L-alanine residue are connected with a hydrophobic double-chain segment by a tertiary amide linkage and with a connector unit to a polar head moiety by a secondary amide linkage, respectively. Thus, the morphological stability comes from the formation of efficient hydrogen-belt domains through intramembrane hydrogen-bonding interactions among amino acid residues. Characteristic morphological features for the peptide lipids and their aggregation behavior have been described elsewhere [23].

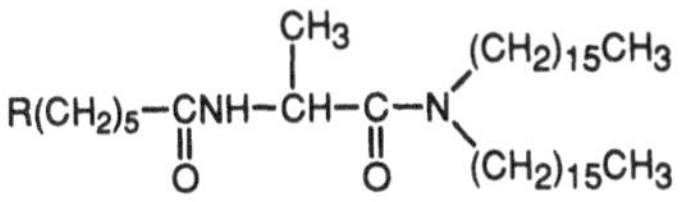

$$R(CH_2)_5-\underset{\underset{O}{\|}}{C}NH-\underset{\underset{CH_3}{|}}{C}H-\underset{\underset{O}{\|}}{C}-N\underset{(CH_2)_{15}CH_3}{\overset{(CH_2)_{15}CH_3}{<}}$$

9 : R = (SO$_3^-$)

10 : R = (CH$_3$)$_3$N$^+$

3 Molecular Recognition Behavior of Artificial Receptors in Aqueous Solution

3.1 Guest Binding Ability

Binding constants (K) for the 1:1 host–guest complexes of artificial receptors with a hydrophobic and anionic guest, ANS (**11**), were evaluated in aqueous media by fluorescence spectroscopy as listed in Table 1 [17, 37, 39, 47]. The K values for the octopus, the steroid, and the cage-type cyclophanes are much greater than the corresponding values for simple macrocycles: 5.5×10^2 for the N,N',N'',N'''-tetramethyl derivative of **1** at pH 2 [50], and 6.6×10^3 dm^3 mol^{-1} for **2** at pH 1.95 [51]. The ANS molecule is incorporated into the three-dimensional inner cavity provided by each host molecule, as confirmed by ^{1}H NMR spectroscopy. Large K values were also observed for these receptors with another anionic guest, TNS (**12**), and various nonionic ones. On the other hand, these hosts exhibit no capacity of binding cationic guest molecules. Thus, it becomes common to see that potent molecular discrimination by the three-dimensionally extended hydrophobic cavity is performed on the basis of the electrostatic interaction in aqueous media, even though the hydrophobic effect is crucial for guest-binding.

Unique inclusion behavior in reflection of the induced-fit binding mechanism is observed when an organic stock solution of octopus cyclophane **3** is injected into an aqueous medium containing ANS for the host–guest complexation study [17]. A circular dichroism (CD) spectrum does not undergo any change for **3** upon complexation with ANS, indicating that the conformation around L-

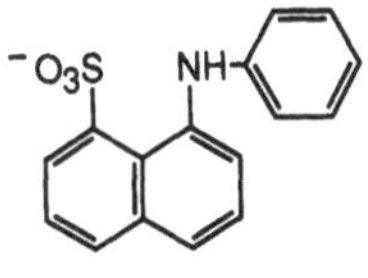

11

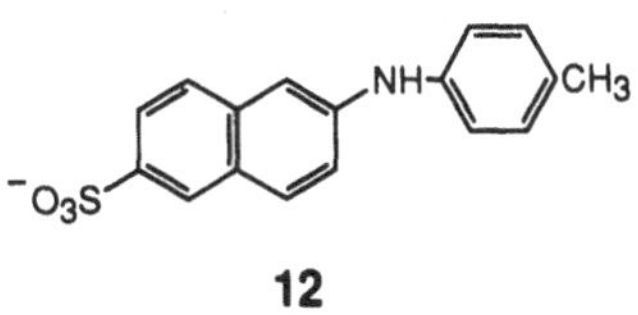

12

Table 1. Binding constants (K) for host–guest complexes of artificial receptors with **11**, and microenvironmental polarity parameters (E_T^N) and steady-state fluorescence polarization values (P) for the guest bound to hosts in aqueous solution at 30.0 °C[a]

Receptor	$K/dm^3 mol^{-1}$	E_T^N	P
3	5.3×10^5	0.57	0.33
4	1.5×10^6	0.58	0.29
5	3.3×10^5	0.41	0.14
6	1.1×10^5	0.64	0.07
7	2.8×10^4	0.23	0.40

[a] In phosphate buffer (10 mmol dm^{-3}) at pH 8.0 for **3**, **4**, and **5**; in acetate buffer (10 mmol dm^{-3}) at pH 5.0 and 4.1 for **6** and **7**, respectively (Cited from [17, 37, 39, 47])

aspartate residues of **3** is fixed regardless of the complexation mode. On the other hand, the fluorescence intensity originated from ANS shows a biphasic time course upon addition of an organic solution of **3** to an aqueous buffer solution of ANS. The fluorescence maximum also undergoes a time-dependent biphasic blue shift. Such behavior is consistent with fast incorporation of the guest molecule into the hydrophobic host cavity followed by slow and long-range conformational change of the host, as induced by the incorporated guest. At the final stage of this binding process, a highly desolvated microenvironment is apparently provided by the host cavity so that the tight host–guest interaction becomes effective. This complexation behavior is also clearly reflected in the steady-state fluorescence polarization parameter (P). Thus, the biphasic host–guest interaction characteristic of host **3** is derived from rigid conformations of the macrocyclic skeleton and the L-aspartate moiety, presumably caused by intramolecular hydrogen bonding that is pronounced most in organic solvents. Such a hydrogen bonding interaction must be retained even after an organic stock soluton is injected into an aqueous buffer. On the other hand, the biphasic behavior is not observed for host–guest complexation of octopus cyclophane **4** with ANS, suggesting that the tetraaza[6.1.6.1]paracyclophane skeleton is more flexible than the tetraaza[3.3.3.3]paracyclophane ring.

The cage-type peptide cyclophanes (**7** and **8**) exhibit discrimination toward steroid hormones, as effected by hydrophobic and π-π interactions. In addition, the chirality-based discrimination between α- and β-estradiol as well as between D- and L-amino acids bearing an aromatic moiety is performed on the basis of their capacity of forming efficient hydrogen bonding with the host molecules in aqueous media [41, 43].

3.2 Microenvironments in Large Hydrophobic Cavities

The microscopic polarity [52] experienced by ANS at the binding site provided by each host was evaluated from the fluorescence maximum observed for the guest. As listed in Table 1, the steroid cyclophane is able to incorporate one ANS molecule into its three-dimensionally extended hydrophobic cavity created by the four steroid moieties and the macrocyclic skeleton in a manner as performed by the octopus cyclophanes. It is noteworthy, however, that the specific structural rigidity of the steroid moieties is reflected in the P value of the guest; the P values in the cavities of **5** and **6** are smaller than those in cavities of the octopus cyclophanes (Table 1). The rotational correlation times (θ) were evaluated from the observed P values and fluorescence lifetimes on the basis of the Perrin's equation [53]: $\theta = 23.0$, 24.4, 5.7, and 2.3 ns for **3, 4, 5,** and **6**, respectively, at 30.0 °C. This means the flexible hydrocarbon chains of the octopus cyclophane are capable of grasping the guest more tightly than the rigid hydrophobic moieties of the steroid cyclophane. It must be noted that the guest-binding site of cage-type cyclophane **7** is much more desolvated and rigid than those of the octopus and steroid cyclophanes in the light of the corresponding microenvironmental parameters.

4 Constitution of Supramolecular Assemblies

A biomembrane is an excellent example of supramolecular assemblies, in which various functional molecules are structurally organized for molecular recognition. In order to develop artificial supramolecular systems capable of mimicking biomembrane functions, it seems important to investigate molecular recognition by macrocyclic hosts embedded in synthetic bilayer membranes.

We have previously clarified that ionic peptide lipids generally undergo aggregation in aqueous media to give bilayer aggregates of multiwalled vesicles and/or lamellae in the dispersion state and single-walled vesicles or small particles upon sonication of the dispersion samples [54–57]. The physical shapes and aggregation modes of these aggregates are nearly identical with those of vesicles formed with naturally occurring phospholipids having aliphatic double-chains of comparable size. However, the single-walled vesicles formed with the present lipids are morphologically much more stable than those with the latter and stay in solution over a prolonged period of time without meaningful morphological changes.

We employed here an anionic peptide lipid (**9**) to form hybrid assemblies with artificial receptors; the octopus, the steroid, and the cage-type peptide cyclophanes. Amphiphile **9** affords multilayered aggregates, bent lamellae and vesicles, when dispersed in aqueous media as confirmed by negative staining electron microscopy [57]. Sonication of an aqueous dispersion sample of **9** with a probe-type sonicator gives a clear solution, and its electron micrograph shows the presence of small particles in the diameter range 20–50 nm for which internal aqueous compartments are not identified clearly. A similar aggregate morphology was also confirmed for a cationic peptide lipid (**10**) in the aqueous dispersion state and in its sonicated solution [56].

We then investigated the formation of hybrid molecular assemblies in combinations of anionic peptide lipid **9** with cage-type hosts **7** and **8** after a previous method [44]. Lamella-type aggregates are observed for a mixture of host **7** and lipid **9** at a 1:20 molar ratio in the dispersion state by negative staining electron microscopy. Phase transition parameters (temperature at peak maximum, T_m; enthalpy change, ΔH; entropy change, ΔS; half-width of an endothermic peak, $\Delta T_{1/2}$) and hydrodynamic diameters (d_{hy}) for the bilayer aggregates in the presence and absence of the macrocycles were evaluated by means of ultrasensitive differential scanning calorimetry (DSC) and dynamic light-scattering (DLS) measurements, respectively. The DSC thermograms for aqueous dispersion samples of lipid **9** in the presence and absence of the cage-type peptide cyclophanes are shown in Fig. 4, and the phase transition parameters are listed together with the d_{hy} values in Table 2. The T_m and the d_{hy} values for the dispersion samples do not undergo significant change even in the presence of the cyclophane host. However, other phase-transition parameters (ΔH, ΔS, and $\Delta T_{1/2}$) are subjected to changes, reflecting formation of the hybrid assemblies: 4–18% decrease in ΔH, 5–18% decrease in ΔS, 110–120% increase in $\Delta T_{1/2}$.

Hybrid molecular assemblies, each composed of the cage-type peptide cyclophane and lipid **9**, are subjected to significant changes as regards aggregate morphology upon sonication with a probe-type sonicator at 30 W power for 15 min. The endothermic peak is broadened and somewhat shifted to a lower tempeature range and the d_{hy} value for the sonicated solution is decreased, suggesting that multi-walled aggregates are transformed into smaller vesicles. In addition, the electron micrographs show the presence of small particles with a diameter of *ca.* 100 nm. Incorporation of cage-type hosts **7** and **8** into the bilayer

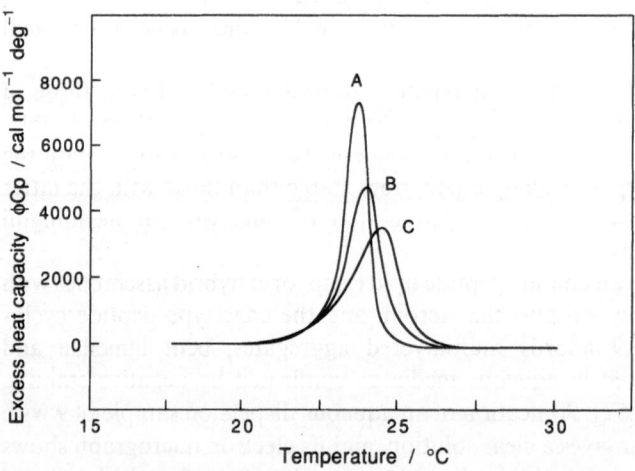

Fig. 4. DSC thermograms for aqueous dispersions of peptide lipid **9** $(1.0 \times 10^{-3} \text{ mol dm}^{-3})$ in aqueous phosphate buffer $(10 \text{ mmol dm}^{-3})$ at pH 7.0: **A** in the absence of cyclophane; **B** in the presence of cage-type cyclophane **7** $(5.0 \times 10^{-5} \text{ mol dm}^{-3})$; **C** in the presence of cage-type cyclophane **8** $(5.0 \times 10^{-5} \text{ mol dm}^{-3})$

Table 2. Phase transition parameters for aggregates in the dispersion state[a]

Aggregate[b]	$T_m/°C$	$\Delta H/\text{kJ mol}^{-1}$	$\Delta S/\text{J K}^{-1}\text{mol}^{-1}$	$\Delta T_{1/2}/°C$	$d_{hy}{}^c/\text{nm}$
9	23.7	31.9	108	0.80	290
7 + 9	23.9	30.5	103	1.65	290
8 + 9	24.4	26.3	88	1.75	280

[a] In phosphate buffer (10 mmol dm^{-3}) at pH 7.0
[b] Concentrations in mol dm^{-3}: lipid, 1.0×10^{-3}, receptors, 2.0×10^{-5}
[c] At 30.0 °C (Cited from [44])

vesicle was more directly characterized by gel-filtration chromatography on a column of Sephadex G-50 with aqueous phosphate buffer as eluent at 30.0 °C; both hosts were eluted together with vesicles at the void volume of the column (Fig. 5). The amount of the incorporated host in the vesicle was determined by electronic spectroscopy. The result reveals that both of the nonionic 7, having only a very limited solubility in water, and the water soluble cationic 8 are fully incorporated into the bilayer vesicle. Moreover, similar gel-filtration chromatograms were obtained for both hosts at 10 °C. This indicates that hosts 7 and 8 remain in the interior vesicle in the gel state and never diffuse out from it. Therefore, the favorable formation of such hybrid assemblies is attributed not only to the hydrophobicity of the host molecules but also to an efficient electrostatic interaction between the cationic host and the anionic polar head group of amphiphile 9 in the aggregated state.

Both 7 and 8 were observed to show an intense absorption band: λ_{max} at 209 nm ($\varepsilon = 2.4 \times 10^4\ \text{dm}^3\,\text{mol}^{-1}\text{cm}^{-1}$) for 7 and λ_{max} at 229 nm ($\varepsilon = 2.7 \times 10^4\ \text{dm}^3\,\text{mol}^{-1}\,\text{cm}^{-1}$) for 8 in aqueous media, respectively. These strong absorption bands seem to originate from an electric transition along the long molecular axis of the bridging segment involving a pyridyl or pyridinium moiety, since the transition energy is appreciably sensitive to the solvent nature: the λ_{max} value for 7 ranges from 209 nm in water to 253 nm in dioxane, whereas the λ_{max} value for 8 ranges from 229 nm in water to 247 nm in dioxane. The microenvironmental polarities experienced by 7 and 8 in the bilayer vesicle, which is in both gel and liquid crystalline states, are evaluated on the basis of a correlation between λ_{max} and solvent polarity parameter. The microenvironmental polarities around the hosts in the liquid-crystalline bilayer vesicle are equivalent to those provided by formamide [E_T^N, 0.799] and p-cresol [E_T^N, 0.929] for 7 and 8, respectively. The E_T^N values for the hosts in the gel-state vesicle are somewhat larger [$E_T^N = 0.88$ and 0.96 for 7 and 8, respectively]. In any cases, 7 and 8 are considered to be placed in a relatively polar domain close to the vesicular surface so that some of the molecular holes of these hosts are exposed to the bulk aqueous phase and are ready to incorporate guest molecules into their internal cavities.

Formation of the hybrid assemblies of peptide lipid 9 with the octopus cyclophanes (3 and 4) and the steroid cyclophanes (5 and 6) was also character-

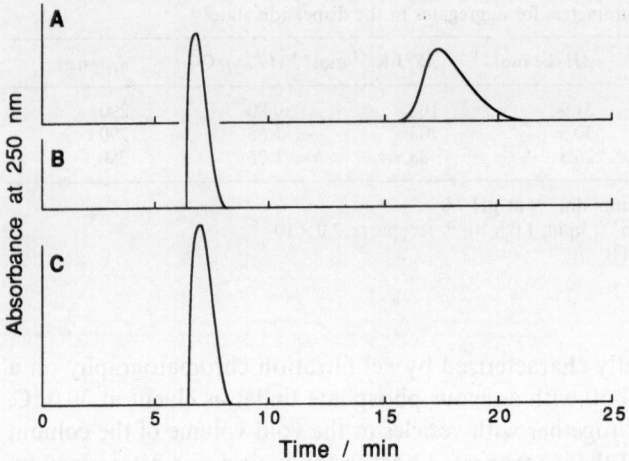

Fig. 5A–C. Gel-filtration chromatograms on a column (column size, 10 mm (I.D.) × 200 mm; flow rate, 2.0 cm³ min⁻¹) of Sephadex G-50 with aqueous phosphate buffer (10 mmol dm⁻³) at pH 7.0 as eluent: **A** cage-type cyclophane **8** alone (1.0×10^{-7} mol); **B** hybrid assembly formed with **7** (1.0×10^{-7} mol) and **9** (2.0×10^{-6} mol); **C** hybrid assembly formed with **8** (1.0×10^{-7} mol) and **9** (2.0×10^{-6} mol)

ized by means of ultrasensitive DSC and DLS measurements. On addition of each host to an aqueous dispersion of **9** in a molar ratio of 1:40, there was observed little change in the T_m value in a manner analogous to those for the hybrid assemblies of lipid **9** with the cage-type cyclophanes (**7** and **8**). On the other hand, phase-transition and hydrodynamic parameters undergo some changes; 3–23% decrease in ΔH, 4–23% decrease in ΔS, 18–33% increase in $\Delta T_{1/2}$, and 21–26% decrease in d_{hy}, reflecting formation of the hybrid assemblies [37]. When a steroid derivative, the monomeric analogue of **5** with respect to the steroid fragment, is added to the aqueous dispersion of **9** in a 1:10 molar ratio, the phase transition parameters and the d_{hy} value are affected to much lesser extents, reflecting its weaker perturbation effect on the bilayer membrane structure.

5 Supramolecular Effects in Molecular Recognition by Hybrid Assemblies

5.1 Octopus and Steroid Cyclophanes Embedded in Synthetic Bilayer Membranes

ANS (**11**) and TNS (**12**) are well known fluorescent probes whose emission is extremely sensitive to changes in the environmental properties around the molecules [58, 59]. Thus, these fluorescent probes were employed to evaluate molecular recognition behavior of hybrid assemblies formed with the macrocyclic hosts (3, 4, 5, and 6) and the peptide lipid [37, 39]. Figure 6A shows fluorescence spectra of ANS in an aqueous solution of 9 with and without steroid cyclophane 5. The solution was sonicated with a probe-type sonicator at 30 W power for 1 min to obtain the bilayer vesicle. While the anionic ANS alone interacts weakly with the anionic lipid aggregate, the hybrid assembly formed with 5 and 9 strongly incorporates ANS into its hydrophobic domain. Such a drastic microenvironmental change is evidently experienced by the guest on the basis of the experimental facts; an increase in the fluorescence intensity and a blue-shift of the fluorescence maximum. An analogous spectral change is observed for an aqueous dispersion of the hybrid assembly, although the background intensity is raised owing to light scattering from the aggregates. Another guest TNS is also effectively bound to the identical hybrid assembly, as shown in Fig. 6B.

Each of the hybrid systems, composed of steroid cyclophane 6 and octopus cyclophanes 3 and 4 in combination with lipid 9, also exhibits effective guest-binding in comparison with the cyclophane-free system as reflected in the fluorescence features of ANS and TNS, which are similar to those shown in Fig. 6. On the other hand, the hybrid assembly formed with a steroid derivative, the monomeric analogue of 5 with respect to the steroid fragment, and lipid 9 in a molar ratio of 1:10 does not enhance the fluorescence intensities of the guests to any significant extent.

The results may provide a guidepost for designing hybrid assemblies which are capable of performing effective binding of anionic and hydrophobic guests that are not incorporated into the anionic bilayer membrane itself; macrocyclic

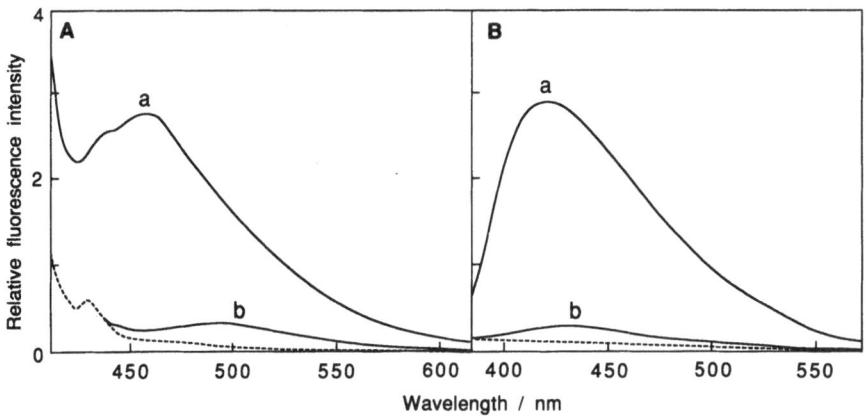

Fig. 6A,B. Fluorescence spectra: **A** 11 in a sonicated solution of 9; **B** 12 in an aqueous dispersion of 9, with (a) and without (b) 5 in aqueous phosphate buffer (10 mmol dm^{-3}) at pH 8.0 and 20.0 °C. Concentrations in mol dm^{-3}: guests, 1.0×10^{-6}; 9, 4.0×10^{-4}; 5, 1.0×10^{-5}. The dotted lines refer to background spectra originating from 9 alone (Taken from [37])

hosts with polycationic charges provide effective binding sites in the anionic bilayer membrane. Unfortunately, the guest-binding constants were not obtained because the light-scattering phenomena caused by these aggregates interfered with the exact evaluation of the fluorescence intensities of the guests.

The microenvironmental polarity parameters for ANS and TNS bound to various hosts are listed in Table 3. These values are independent of temperature over a range of 10–40 °C. In the absence of any macrocyclic hosts, ANS is bound to the membrane in its surface domain while TNS to the hydrogen-belt domain [60, 61] interposed between the polar surface region and the hydrophobic domain composed of double-chain segments in the light of the E_T^N values; the microenvironments for the former and the latter are close to that provided by water ($E_T^N = 1.000$) and equivalent to that in ethanol ($E_T^N = 0.654$), respectively. Such a difference in microenvironmental polarity presumably comes from the difference in molecular shape; TNS is more slender than ANS.

The steroid cyclophanes provide less polar microenvironments for ANS and TNS by forming the hybrid assemblies with the peptide lipid. To our surprise, the microenvironment around the ANS molecule incorporated into the hybrid assembly formed with lipid **9** and steroid cyclophane **5** is equivalent to that provided by hexane ($E_T^N = 0.009$). In contrast, the microscopic polarity experienced by TNS in the identical hybrid assembly is as polar as 1-pentanol ($E_T^N = 0.568$). On the other hand, the hybrid assembly formed with lipid **9** and steroid cyclophane **6** binds the TNS molecule in a less polar microenvironment than ANS. Meanwhile, both of the octopus cyclophanes and their hybrid assemblies formed with the peptide lipid incorporate both guest molecules into the comparable binding sites with respect to the microenvironmental polarity.

In order to evaluate microscopic viscosity around the guest incorporated into the hybrid assemblies, the steady-state fluorescence polarization measurements were performed for ANS and TNS (Table 3). The P value for ANS

Table 3. Microenvironmental polarity parameters (E_T^N) and steady-state fluorescence polarization values (P) for guests bound to hybrid assemblies formed with artificial receptors and peptide lipid in aqueous solution at 30.0 °C[a]

Hybrid[b]	11		12	
	E_T^N	P	E_T^N	P
9	0.90	0.16	0.66	0.24
3 + 9	0.57	0.28	0.63	0.30
4 + 9	0.57	0.30	0.58	0.29
5 + 9	0.01	0.33	0.58	0.33
6 + 9	0.59	0.33	0.24	0.35

[a] In phosphate buffer (10 mmol dm^{-3}) at pH 8.0 for **3**, **4**, and **5**; in acetate buffer (10 mmol dm^{-3}) at pH 5.0 for **6**
[b] Concentrations in mol dm^{-3}; lipid, 4.0×10^{-4}; receptors, 1.0×10^{-5}; guests, 1.0×10^{-6} (Cited from [37, 39])

incorporated into steroid cyclophane **5** in aqueous solution remains nearly constant at 0.14 in over a range of 10–40 °C. In the case of the corresponding hybrid assembly formed with the peptide lipid, however, the P value decreases significantly as temperature is raised along with a slight inflection in the phase transition temperature range. Hence, the ANS molecule is obviously incorporated into the hydrophobic domain of the aggregate and its molecular motion is subjected to change by the phase transition. The P value for ANS bound to octopus cyclophane **4** in aqueous solution decreases monotonously as temperature is raised. Although the P vs temperature correlation line for the hybrid system composed of **4** and **9** is not much separated from that for **4** alone in aqueous solution, the P value for the hybrid assembly is subjected to change by temperature in a biphasic manner, exhibiting a slight inflection in the T_m range. It must be noted that similar correlations between temperature and P were observed for TNS with and without the bilayer membrane.

Since the P value is subject to change by the fluorescence lifetime (τ) and the rotational correlation time (θ), these values for ANS bound to the hosts were evaluated in the presence of the bilayer membrane. All the τ values for ANS bound to the hosts are large ($\tau = 10$–15 ns) relative to that in water ($\tau = 0.55$ ns). This means that the guest molecules are placed in hydrophobic microenvironments well separated from the bulk aqueous phase. It is noteworthy that the θ values for ANS in the hybrid assemblies with the steroid cyclophanes are much larger than those in the hybrid assemblies formed with the octopus cyclophanes. In the liquid-crystalline membrane at 30.0 °C, the θ values are 17.8, 23.0, 40.5, and 39.9 ns for the hybrid assemblies of **9** with **3**, **4**, **5**, and **6**, respectively. Moreover, the θ values for ANS increase significantly in the gel-state membrane: $\theta = 31.6$ and 66.7 ns at 20.0 °C for octopus host **3** and steroid host **5**, respectively. The θ values for ANS bound to liposomal membranes formed with lecithin and those bound to various proteins are in ranges 3–6 and 9–63 ns, respectively [58]. Hence, the θ value for ANS incorporated into the hybrid assembly in the gel state, which is formed with cyclophane **5** and lipid **9**, seems to be the largest one for ANS so far observed. It must be pointed out that such remarkable restriction of the molecular motion is performed by a combination of the steroid cyclophane, which is incapable of performing tight guest binding by itself, with the anionic peptide lipid, which cannot bind an anionic guest effectively in its aggregated state. In contrast, the molecular motion of ANS bound to the octopus cyclophanes is not subjected to significant change in their hybrid assemblies formed with the peptide lipid.

On the above ground, guest binding modes of the present hybrid assemblies are classified into two types as schematically shown in Fig. 7; a guest molecule is included in the proximity of the hydrogen-belt domain of the bilayer membrane on the one hand, and is bound to the hydrophobic region composed of the hydrocarbon double chains on the other. The hybrid assemblies formed with the octopus cyclophane and the peptide lipid exercise the former binding mode toward both ANS and TNS (Fig. 7A). Even though the macrocyclic skeletons of octopus cyclophanes **3** and **4** are different from each other, such structural

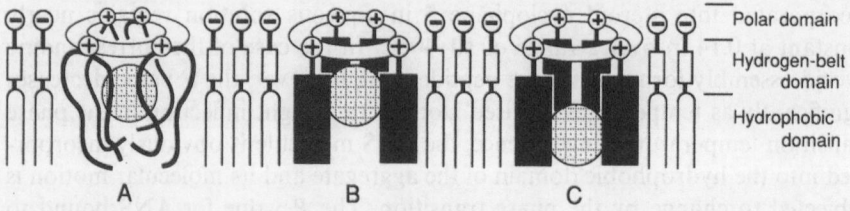

Fig. 7A–C. Schematic representations for guest-binding modes of hybrid assemblies formed with peptide lipid **9** and octopus (**3** and **4**) and steroid (**5** and **6**) cyclophanes: a guest molecule is located in the hydrogen-belt domain (**A** for **11** and **12** each bound to **3** and **4**; **B** for **11** and **12** bound to **6** and **5**, respectively); a guest molecule is located in the hydrophobic domain (**C** for **11** and **12** bound to **5** and **6**, respectively)

difference scarcely affects the guest-binding behavior. On the other hand, the hybrid assemblies formed with the steroid cyclophanes and the peptide lipid exercise the former and the latter binding modes toward TNS and ANS, depending on the structural characteristics of the host molecules (Fig. 7B, 7C). Such site-directed binding of anionic guest molecules to the host cavity embedded in the bilayer membrane cannot be performed by the hybrid assemblies formed with cationic lipid **10** and the present cationic cyclophane hosts.

5.2 Cage-Type Cyclophanes Embedded in Synthetic Bilayer Membranes

A CD spectroscopic study gave us important information as regards the chiral properties of the cage-type cyclophane cavities [41, 43–45]. Cage-type host **8** shows a CD band at 242 nm with a molecular ellipticity ($[\theta]$, deg cm^2 dmol^{-1}) of $+1.0 \times 10^5$ in aqueous HEPES buffer at 30.0 °C, while a peptide cyclophane bearing L-valine residues, one of the precursors of host **7**, and a bridging component analogue of **8** do not show any detectable CD bands over a relatively wide wavelength range. These results suggest that the four pyridinium moieties bound to the chiral L-valine residues in the bridging segments of **8** approach close to each other and are twisted in the same direction. Such a helical conformation for **8** seems to be caused by the chiral nature of the optically active valine residues in the bridging segments in the light of minimum energy conformation for the host in the gas phase, as obtained on the basis of molecular mechanics and dynamics calculations.

Furthermore, a cage-type peptide cyclophane bearing D-valine residues in place of L-valine moieties shows a CD band at 244 nm with $[\theta]$ of -1.1×10^5 in HEPES buffer at 30.0 °C. It becomes apparent that the four pyridinium moieties bound to the chiral D-valine residues are similarly twisted in the same direction, but the twisted direction of bridging components in the host is opposite to that evaluated for the host bearing L-valine residues. A similar asymmetric character of the internal cavities provided by the nonionic cage-type cyclophanes bearing L- and D-valine residues was also evidenced by the identical methods.

The CD band intensity originated from the host (8) is enhanced in the bilayer vesicle ($[\theta]$ $+1.5 \times 10^5 \deg \text{cm}^2 \text{dmol}^{-1}$, at 247 nm) relative to the corresponding value in the HEPES buffer at 30.0 °C. In addition, no CD spectral change was observed for at least one day at 30.0 °C. This result indicates that the four pyridinium moieties bound to the chiral L-valine residues in 8 assume highly restricted conformations in the bilayer membrane. Thus, the hybrid assembly seems to furnish a chiral guest-binding site different from that provided by 8 alone in aqueous media without the vesicle.

Under such circumstances, the guest-binding behavior of the hybrid assemblies toward various hydrophobic molecules was examined by changing the guest concentration, while the host concentration was maintained constant in aqueous phosphate buffer. The following dyes were adopted as guest molecules: Naphthol Yellow S (13), Dimethylsulfonazo III (14), Bromopyrogallol Red (15), and Orange G (16). The CD band at around 247 nm is weakened in intensity as the guest concentration is increased (Fig. 8). The decrease in CD band intensity at 247 nm is attributable to conformational changes around the pyridinium moieties of 8 upon complexation with the guest molecules to form thermodynamically stable complexes. The molecular mechanics and dynamics calculations reveal that the pyridinium moieties in the bridging segments of 8 are separated from each other as the guest molecule is incorporated into the internal cavity of the host (Figs. 9 and 10).

The stoichiometry for the complexes formed with host 8 embedded in the bilayer membrane and the guest was investigated by the molar ratio method [62]. A plot of the $[\theta]$ value for 8 against the concentration of Orange G reveals that 8 embedded in the bilayer vesicle forms a complex with Orange G in a 1:1 molar ratio. The same 1:1 stoichiometry was confirmed for other complexes. The formation constants (K) for the 1:1 host–guest complexes in the bilayer

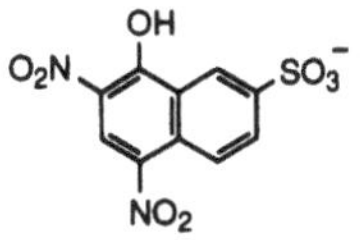

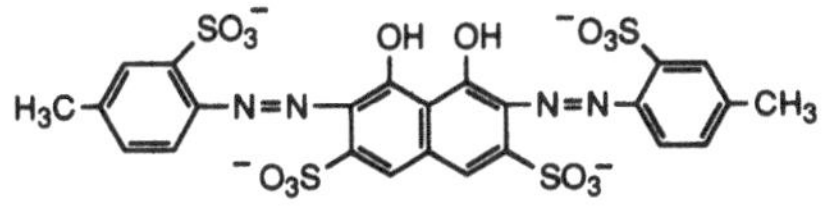

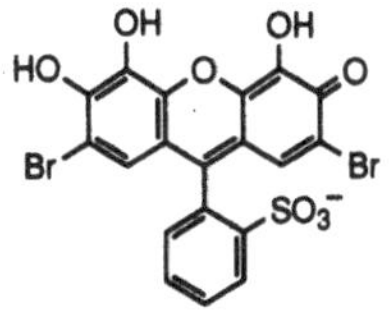

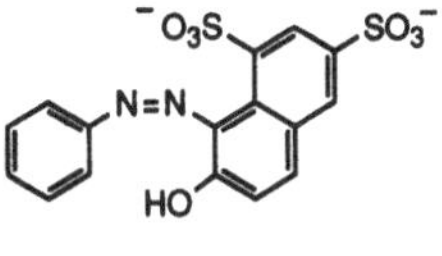

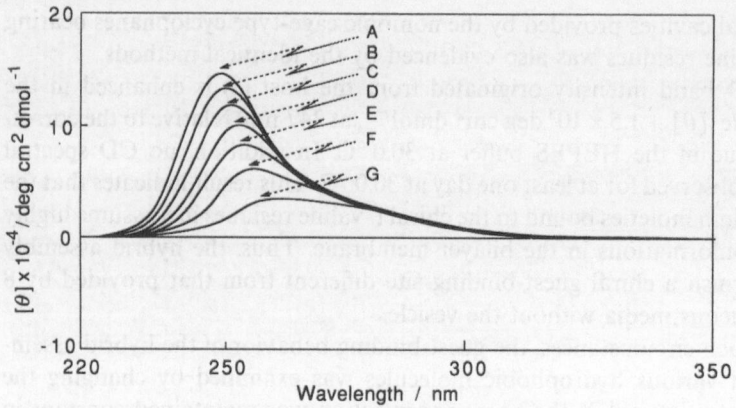

Fig. 8. CD spectral change for hybrid assembly formed with **8** (5.0×10^{-5} mol dm^{-3}) and **9** (1.0×10^{-3} mol dm^{-3}) upon addition of Dimethylsulfonazo III in aqueous phosphate buffer (10 mmol dm^{-3}) at pH 7.0 and 30.0 °C. Concentrations of the guest in μ mol dm^{-3}: A, 0; B, 5.0; C, 20; D, 35; E, 50; F, 75; G, 100 (Taken from [44])

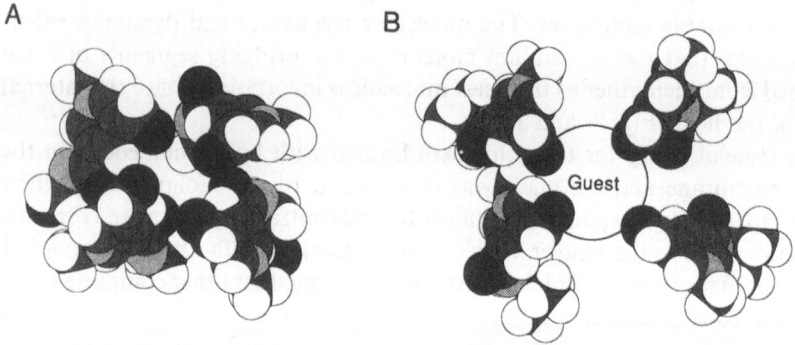

Fig. 9A,B. Space-filling models optimized conformationally: **A** for **8**; **B** for a complex of **8** with Naphthol Yellow S. Two macrocyclic rings of the host are removed for simplicity to show conformational change in the binding moieties of **8** (Taken from [44])

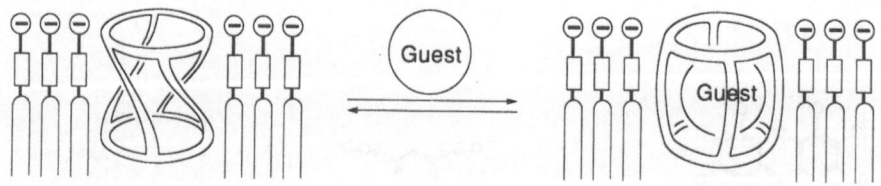

Fig. 10. Schematic representation of guest-binding behavior of hybrid assemblies formed with peptide lipid **9** and cage-type cyclophanes **7** and **8**, indicating conformational change of the host cavity

membrane were evaluated on the basis of the CD spectral changes. As is apparent from the formation constants listed in Table 4, the binding affinities of cage-type hosts **7** and **8** toward guest molecules are effectively retained even when the hosts are incorporated into the bilayer vesicle. It must be noted that the CD spectra for the hybrid assemblies, observed in the presence of a sufficient excess of the guest, are similar to those for the corresponding hosts in the absence of the vesicle under otherwise identical conditions. The result indicates that the host molecules located in both inner and outer monolayers of the membrane are equally capable of binding the anionic guest. This leads us to conclude that both hosts are able to transfer the anionic guest from the bulk aqueous phase to the inner membrane monolayer, acting as a functionalized channel or carrier.

The chirality-based molecular discrimination behavior of the hybrid assembly formed with **7** and **9** toward optically active α-amino acids was examined by means of ^{1}H NMR spectroscopy in D_2O at 30.0 °C [63]. When the bilayer vesicle of lipid **9** was added to aqueous solutions containing L- and D-tryptophan individually, any significant chemical shift changes were not observed for the guest molecules. This indicates that interactions between the guest molecules and the anionic vesicular surface are very weak. On the other hand, all the ^{1}H NMR signals due to the guests in the hybrid assemblies were broadened and underwent substantial upfield shifts, except for the signal due to H-2 proton on the indole ring that shifted to downfield. It is clear now that the guest molecule is incorporated into the three-dimensional hydrophobic cavity of the cage-type cyclophane embedded in the bilayer membrane. The binding constants (K) for the 1:1 host–guest complexation were evaluated on the basis of the computer-aided least-squares curve fitting method applied to the NMR titration data: $K = 5 \times 10^4$ and $3 \times 10^4 \, dm^3 \, mol^{-1}$ for L-tryptophan and D-tryptophan, respectively. Since the present hybrid assembly shows no capacity of binding aliphatic guests, such as L-methionine and L-threonine, the hybrid assembly formed with host **7** and lipid **9** seems to perform chiral recognition toward hydrophobic and aromatic α-amino acids through hydrophobic and π-π interactions.

Table 4. Binding constants (K) for host–guest complexes formed in aqueous solution and bilayer membrane at 30.0 °C[a]

Guest[b]	Aqueous solution		Bilayer membrane	
	7	8	7 + 9	8 + 9
13	5.5×10^4	3.7×10^5	2.7×10^4	4.1×10^4
14	3.8×10^4	3.4×10^5	3.0×10^4	5.0×10^4
15	5.8×10^4	2.5×10^5	2.1×10^4	3.3×10^4
16		1.2×10^5		2.6×10^4

[a] In acetate buffer (10 mmol dm^{-3}) at pH 4.1 for **7**; in phosphate buffer (10 mmol dm^{-3}) at pH 8.0 for **8**, **7 + 9**, and **8 + 9**
[b] Concentrations in mol dm^{-3}: lipid, 1.0×10^{-3}; receptors, 5.0×10^{-5}; guests, $5.0 \times 10^6 - 2.0 \times 10^{-4}$ (Cited from [44])

Y. Murakami et al.

6 Concluding Remarks

Molecular recognition is one of the most important subjects in supramolecular chemistry. We have described our recent approaches to the formation of large hydrophobic cavities capable of providing effective guest-binding sites in bilayer membranes as well as in aqueous solution. Three types of artificial receptors, each having a macrocyclic skeleton (or skeletons) as a basic structural component, were designed and synthesized: octopus cyclophanes, steroid cyclophanes, and cage-type cyclophanes.

The hydrophobic cavity provided by the *octopus cyclophane* is significantly flexible, and capable of performing molecular recognition toward hydrophobic guest molecules of various bulkiness through the induced-fit mechanism, not only in aqueous solution but in the bilayer membrane formed with a synthetic peptide lipid. However, such a flexible hydrophobic cavity seems to be unfavorable for size-sensitive molecular discrimination.

The *steroid cyclophane* also provides a sizable and well-desolvated hydrophobic cavity in aqueous media in a manner as observed for the octopus cyclophane. The molecular recognition ability of the steroid cyclophane is inferior to that of the octopus cyclophane in aqueous solution due to the structural rigidity of steroid segments of the former host. When the steroid cyclophane is embedded in the bilayer membrane to form a hybrid assembly, however, the steroid cyclophane becomes superior to the octopus cyclophane with respect to functions as an artificial cell-surface receptor, performing marked guest discrimination.

The *cage-type cyclophane* furnishes a hydrophobic internal cavity for inclusion of guest molecules and exercises marked chiral discrimination in aqueous media. The host embedded in the bilayer membrane is capable of performing effective molecular recognition as an artificial cell-surface receptor to an extent comparable to that demonstarated by the host alone in aqueous media.

To the best of our knowledge, the present findings can be cited as the first successful examples of hybrid assemblies that are eligible for mimicking sensor functions of biological receptors embedded in biomembranes. We believe that the present study provides a useful guidepost for designing functionalized cyclophane hosts as well as hybrid assemblies for functional simulation of naturally occurring receptors.

Acknowledgements. We are grateful to many collaborators and students, whose names appear in the references cited herein, for their fruitful contributions to the work described here. The present work has been supported primarily by a Special Distinguished Grant for Scientific Research No. 02102006 from the Ministry of Education, Science, and Culture of Japan.

7 References

1. Ehrlich P (1906) Proc Roy Soc London (Biol) 66: 424
2. Langley JN (1906) J Physiol (London) 33: 374
3. Alberts B, Bray D, Lewis J, Raff M, Roberts K, Watson JD (1989) Molecular biology of the cell, 2nd edn. Garland Publishing, New York, p 681
4. Lehn J-M (1988) Angew Chem Int Ed Engl 27: 89
5. Diederich F (1988) Angew Chem Int Ed Engl 27: 362
6. Koga K, Odashima K (1989) J Incl Phenom 7: 53
7. Murakami Y, Kikuchi J, Hisaeda Y, Ohno T (1991) in: Schneider H-J, Dürr H (eds) Frontiers in supramolecular organic chemistry and photochemistry. VCH, Weinheim, p 145
8. Rebek J Jr (1990) Angew Chem Int Ed Engl 29: 245
9. Schneider H-J (1991) Angew Chem Int Ed Engl 30: 1417
10. Seel C, Vögtle F (1992) Angew Chem Int Ed Engl 31: 528
11. Tabushi I, Kuroda Y, Yokota K (1982) Tetrahedron Lett 23: 4601
12. Lehn J-M (1990) Angew Chem Int Ed Engl 29: 1304
13. Odashima K, Sugawara M, Umezawa Y (1991) Trends Anal Chem 10: 207
14. Murakami Y, Kikuchi J, Ohno T (1990) in : Gokel GW (ed) Advances in supramolecular chemistry, vol 1. JAI Press, Greenwich, p 109
15. Murakami Y, Kikuchi J, Suzuki M, Takaki T (1984) Chem Lett 2139
16. Murakami Y, Kikuchi J, Suzuki M, Matsuura T (1988) J Chem Soc, Perkin Trans 1 1289
17. Murakami Y, Kikuchi J, Ohno T, Hayashida O, Kojima M (1990) J Am Chem Soc 112: 7672
18. Murakami Y, Kikuchi J, Hirayama T (1987) Chem Lett 161
19. Murakami Y, Kikuchi J, Ohno T, Hirayama T (1989) Chem Lett 881
20. Murakami Y, Kikuchi J, Ohno T, Hirayama T, Nishimura H (1989) Chem Lett 1199
21. Murakami Y, Kikuchi J, Ohno T, Hirayama T, Hisaeda Y, Nishimura H (1991) Chem Lett 1657
22. Murakami Y, Kikuchi J, Ohno T, Hirayama T, Hisaeda Y, Nishimura H, Snyder JP, Steliou K (1991) J Am Chem Soc 113: 8229
23. Murakami Y, Kikuchi J (1991) in: Dugas H (ed) Bioorganic chemistry frontiers, vol 2. Springer, Berlin Heidelberg New York, p 73
24. Urushigawa Y, Inazu T, Yoshino T (1971) Bull Chem Soc Jpn 44: 2546
25. Murakami Y, Nakano A, Miyata R, Matsuda Y (1979) J Chem Soc, Perkin Trans 1 1699
26. Murakami Y, Nakano A, Akiyoshi K, Fukuya K (1981) J Chem Soc, Perkin Trans 1 2800
27. Tabushi I, Kimura Y, Yamamura K (1981) J Am Chem Soc 103: 6486
28. Tabushi I, Yamamura K, Nonoguchi H, Hirotsu K, Higuchi T (1984) J Am Chem Soc 106: 2621
29. Kodaka M, Fukaya T (1986) Bull Chem Soc Jpn 59: 2944
30. Lepropre G, Fastrez J (1987) J Incl Phenom 5: 157
31. Odashima K, Itai A, Iitaka Y, Koga K (1980) J Am Chem Soc 102: 2504
32. Takahashi I, Odashima K, Koga K (1984) Tetrahedron Lett 25: 973
33. Breslow R, Czarnik AW, Lauer M, Leppkes R, Winkler J, Zimmerman S (1986) J Am Chem Soc 108: 1969
34. Schneider H-J, Blatter T (1988) Angew Chem Int Ed Engl 27: 1163
35. Lai C-F, Odashima K, Koga K (1989) Chem Pharm Bull 37: 2351
36. Kikuchi J, Egami K, Suehiro K, Murakami Y (1992) Chem Lett 1685
37. Kikuchi J, Matsushima C, Tanaka Y, Hie K, Suehiro K, Hayashida O, Murakami Y (1992) J Phys Org Chem 5: 633
38. Kikuchi J, Matsushima C, Suehiro K, Oda R, Murakami Y (1991) Chem Lett 1807
39. Kikuchi J, Inada M, Miura H, Suehiro K, Hayashida O, Murakami Y (1994) Recl Trav Chim Pays-Bas 113: 216
40. Murakami Y, Kikuchi J, Tenma H (1985) J Chem Soc, Chem Commun 753
41. Murakami Y, Ohno T, Hayashida O, Hisaeda Y (1991) J Chem Soc, Chem Commun 950
42. Murakami Y, Ohno T, Hayashida O, Hisaeda Y (1991) Chem Lett 1595
43. Murakami Y, Hayashida O, Ito T, Hisaeda Y (1992) Chem Lett 497
44. Murakami Y, Hayashida O (1993) Proc Natl Acad Sci USA 90: 1140
45. Murakami Y, Hayashida O, Ito T, Hisaeda Y (1993) Pure Appl Chem 65: 551
46. Murakami Y, Hayashida O, Matsuura S (1993) Recl Trav Chim Pays-Bas 112: 421

Y. Murakami et al.

47. Murakami Y, Hayashida O, Ono K, Hisaeda Y (1993) Pure Appl Chem 65: 2319
48. Mayo L, Olafson BD, Goddard III WA (1990) J Phys Chem 94: 8897
49. Brockerhoff H (1977) in: van Tamelen EE (ed) Bioorganic chemistry, vol 3. Academic Press, New York, p 1
50. Tabushi I, Kuroda Y, Kimura Y (1976) Tetrahedron Lett 3327
51. Odashima K, Soga T, Koga K (1981) Tetrahedron Lett 22: 5311
52. Reichardt C (1988) Solvents and solvent effects in organic chemistry. VCH, Weinheim, p 359
53. Perrin MF (1929) Ann Phys (Leipzig) 26: 169
54. Murakami Y, Nakano A, Fukuya K (1980) J Am Chem Soc 102: 4253
55. Murakami Y, Nakano A, Ikeda H (1982) J Org Chem 47: 2137
56. Murakami Y, Nakano A, Yoshimatsu A, Uchitomi K, Matsuda Y (1984) J Am Chem Soc 106: 3613
57. Murakami Y, Kikuchi J, Takaki T, Uchimura K, Nakano A (1985) J Am Chem Soc 107: 2161
58. Slavik J (1982) Biochim Biophys Acta 694: 1
59. McClure WO, Edelman GM (1966) Biochemistry 5: 1908
60. Murakami Y, Kikuchi J, Nishida K, Nakano A (1984) Bull Chem Soc Jpn 57: 1371
61. Murakami Y, Kikuchi J, Akiyoshi K, Imori T (1986) J Chem Soc, Perkin Trans 2 1445
62. Arimura T, Kawabata H, Matsuda T, Muramatsu T, Satoh H, Fujio K, Manabe O, Shinkai S (1991) J Org Chem 56: 301
63. Murakami Y, Hayashida O, Nagai Y (1994) Recl Trav Chim Pays-Bas 113: 209

Author Index Volumes 151-175

The volume numbers are printed in italics

Dear, K.: Cleaning-up Oxidations with Hydrogen Peroxide. *164*, (1993).

de Mendoza, J., see Seel, C.: *175*, 101-132 (1995).

de Silva, A.P., see Bissell, R.A.: *168*, 223-264 (1993).

Descotes, G.: Synthetic Saccharide Photochemistry. *154*, 39-76 (1990).

Dias, J.R.: A Periodic Table for Benzenoid Hydrocarbons. *153*, 123-144 (1990).

Dietrich-Buchecker, Ch., see Chambron, J.-C.: *165*, 131-162 (1993).

Dohm, J., Vögtle, F.: Synthesis of (Strained) Macrocycles by Sulfone Pyrolysis. *161*, 69-106 (1991).

Dutasta, J.-P., see Collet, A.: *165*, 103-129 (1993).

Eaton, D.F.: Electron Transfer Processes in Imaging. *156*, 199-226 (1990).

El-Basil, S.: Caterpillar (Gutman) Trees in Chemical Graph Theory. *153*, 273-290 (1990).

Fasani, A., see Albini, A.: *168*, 143-173 (1993).

Fontaine, A., Dartyge, E., Itie, J.P., Juchs, A., Polian, A., Tolentino, H., and Tourillon, G.: Time-Resolved X-Ray Absorption Spectroscopy Using an Energy Dispensive Optics: Strengths and Limitations. *151*, 179-203 (1989).

Foote, C.S.: Photophysical and Photochemical Properties of Fullerenes. *169*, 347-364 (1994).

Fossey, J., Sorba, J., and Lefort, D.: Peracide and Free Radicals: A Theoretical and Experimental Approach. *164*, 99-113 (1993).

Fox, M.A.: Photoinduced Electron Transfer in Arranged Media. *159*, 67-102 (1991).

Freeman, P.K., and Hatlevig, S.A.: The Photochemistry of Polyhalocompounds, Dehalogenation by Photoinduced Electron Transfer, New Methods of Toxic Waste Disposal. *168*, 47-91 (1993).

Fuchigami, T.: Electrochemical Reactions of Fluoro Organic Compounds. *170*, 1-38 (1994).

Fuller, W., see Grenall, R.: *151*, 31-59 (1989).

Galán, A., see Seel, C.: *175*, 101-132 (1995).

Gehrke, R.: Research on Synthetic Polymers by Means of Experimental Techniques Employing Synchrotron Radiation. *151*, 111-159 (1989).

Gerratt, J., see Cooper, D.L.: *153*, 41-56 (1990).

Gerwick, W.H., Nagle, D.G., and Proteau, P.J.: Oxylipins from Marine Invertebrates. *167*, 117-180 (1993).

Gigg, J., and Gigg, R.: Synthesis of Glycolipids. *154*, 77-139 (1990).

Gislason, E.A., see Guyon, P.-M.: *151*, 161-178 (1989).

Greenall, R., Fuller, W.: High Angle Fibre Diffraction Studies on Conformational Transitions DNA Using Synchrotron Radiation. *151*, 31-59 (1989).

Gruber, B., see Bley, K.: *166*, 199-233 (1993).

Güdel, H. U., see Colombo, M. G.: *171*, 143-172 (1994).

Gunaratne, H.Q.N., see Bissell, R.A.: *168*, 223-264 (1993).

Guo, X.F., see Zhang, F.J.: *153*, 181-194 (1990).

Gust, D., and Moore, T.A.: Photosynthetic Model Systems. *159*, 103-152 (1991).

Gutman, I.: Topological Properties of Benzenoid Systems. *162*, 1-28 (1992).

Gutman, I.: Total π-Electron Energy of Benzenoid Hydrocarbons. *162*, 29-64 (1992).

Guyon, P.-M., Gislason, E.A.: Use of Synchrotron Radiation to Study-Selected Ion-Molecule Reactions. *151*, 161-178 (1989).

Hadjiarapoglou, L., see Adam, W.: *164*, 45-62 (1993).

Hart, H., see Vinod, T. K.: *172*, 119-178 (1994).

Springer-Verlag
and the Environment

We at Springer-Verlag firmly believe that an international science publisher has a special obligation to the environment, and our corporate policies consistently reflect this conviction.

We also expect our business partners – paper mills, printers, packaging manufacturers, etc. – to commit themselves to using environmentally friendly materials and production processes.

The paper in this book is made from low- or no-chlorine pulp and is acid free, in conformance with international standards for paper permanency.